Windows into the Earth

Aerial view of a rare eruption of Steamboat Geyser in Norris Geyser Basin, July 6, 1984. (Robert B. Smith)

Windows into the Earth

The Geologic Story of Yellowstone and Grand Teton National Parks

Robert B. Smith and Lee J. Siegel

2000

OXFORD
UNIVERSITY PRESS

Oxford New York
Athens Auckland Bangkok Bogotá Buenos Aires Calcutta
Cape Town Chennai Dar es Salaam Delhi Florence Hong Kong Istanbul
Karachi Kuala Lumpur Madrid Melbourne Mexico City Mumbai
Nairobi Paris São Paulo Singapore Taipei Tokyo Toronto Warsaw

and associated companies in
Berlin Ibadan

Published by Oxford University Press, Inc.
198 Madison Avenue, New York, New York 10016

Library of Congress Cataloging-in-Publication Data
Smith, Robert Baer, 1938–
Windows into the earth : the geologic story of Yellowstone and Grand Teton national parks / Robert B. Smith and Lee J. Siegel
p. cm.
Includes bibliographical references and index.
ISBN-13:978-0-19-510597-1 (paper)
1. Geology—Yellowstone National Park.
2. Geology—Wyoming—Grand Teton National Park.
I. Siegel, Lee J. II. Title.
QE79.S58 2000 557.87'5—dc21 99-22287

Book design by Jeff Hoffman

19 18 17 16 15 14 13 12 11 10
Printed in China
on acid-free paper

Dedicated to the memory of Roderick "Rick" Hutchinson, geologist extraordinaire for the National Park Service at Yellowstone National Park. Rick was deeply committed to studying Yellowstone's geology, geysers, earthquakes, and other natural wonders. He was a longtime friend who encouraged and assisted in the preparation of this book. In March 1997, Rick was killed by an avalanche while doing what he loved: conducting fieldwork in the Yellowstone backcountry.

Contents

Preface

Majestic mountains, dazzling arrays of geysers, and magnificent wildlife—bison, bears, elk, and, recently, wolves—are among the world-famous features of Yellowstone and Grand Teton national parks. Yet most visitors are unaware that the crown jewels of the U.S. national park system are the still-evolving products of activity within the Earth—volcanism, earthquakes, movements of tectonic plates, and other geological processes that are key elements in shaping the planet's landscape. Because the geology of Yellowstone and Grand Teton reveals much about these processes, the two parks are truly windows into the Earth.

The forces that sculpted the Teton Range and Jackson Hole, Wyoming, also stretched apart a broad area of the U.S. West known as the Basin and Range Province. Yellowstone, designated the nation's first national park in 1872, was molded both by those forces and by the Yellowstone hotspot, a source of heat and molten rock with roots deep within the Earth. The hotspot produces volcanic activity ranging from hot springs and geysers to cataclysmic eruptions. The beauty of the region helped inspire the global environmental ethic, and its geology has revealed new ideas on volcanism, active mountain building, earthquakes, and glaciation.

Grand Teton and Yellowstone parks together form a "geoecosystem," an idea embracing the concepts of how the ecology of plants and animals depends on their geologic setting: volcanic soils, hydrothermal areas, and climate conditions resulting from mountainous topography built by movements of Earth's crust and the unseen power of the Yellowstone hotspot.

This book looks inside the workings of the Yellowstone hotspot and other geologic forces that have shaped—and continue to shape—the greater Yellowstone–Teton region. We begin with the Hebgen Lake earthquake of 1959—the most violent and deadly geologic disaster in the region in modern time—and end with driving tour guides to the area's geologic wonders.

Salt Lake City, Utah R.B.S.
1999 L.J.S.

Acknowledgments

This book reflects the integration of knowledge from more than 30 years of research in the Yellowstone–Teton region by Bob Smith and his colleagues, and from Lee Siegel's 24-year career writing for the general public about science, particularly geology.

The late Rick Hutchinson of Yellowstone National Park readily gave of his time, accompanied Bob Smith on many field expeditions, and provided photographs for the book. We especially thank Bob Christiansen, Robert Fournier, Dan Dzurisin, and Ken Pierce for discussions on the volcanic history, hydrothermal processes, ground-motion studies, and surface geology of Yellowstone. We are indebted to J. David Love for his many years of encouragement and ideas on the geology of the Teton region. Chuck Meertens and Tony Lowry assisted with Global Positioning System (GPS) satellite measurements of ground movements in the Yellowstone–Teton region. John O. D. Byrd contributed ideas on the evolution of the Teton fault. Art Sylvester worked with Smith on ground-movement studies of the Teton fault. Other co-workers included Harley Benz, Larry Braile, Mike Perkins, Mitch Pitt, Neil Ribe, and John C. Reed, Jr. Dave Drobeck, Erwin McPherson, Sue Nava, Dan Trentman, and Ken Whipp are thanked for their help with the University of Utah earthquake-monitoring efforts.

Preparation of this book included the extended efforts of Mickey Begent for his computer drafting of illustrations. Unless otherwise noted, Henry H. Holdsworth, of the Wild by Nature gallery, provided the excellent photography, including the beautiful photo of elk at Mammoth Hot Springs. Tom Mangelsen, of Images of Nature galleries, loaned us a panoramic camera, provided a magnificent photo of the Tetons, and gave important advice. Chuck Overton rendered the artful portrayals of Yellowstone's giant volcanic eruptions and E. V. Wingert produced a satellite image of the Yellowstone–Teton region. Bill Parry and Ron Bruhn at the University of Utah offered scientific advice and encouragement. Joyce Berry, Helen B. Mules, and Rosemary Wellner edited the book and advised us on publishing issues.

We appreciate the long-term support of Yellowstone National Park personnel Michael Finley, John Varley, John Lounsbury, Stu Coleman, and Anne Deutch. In Grand Teton National Park, we thank Jack Neckels, Bob Schiller, Peter Hayden, Bill Swift, and Marshall Gingery. Hank Harlow, director of the University of Wyoming–National Park Service Research Center in Moran, Wyoming, contributed support and encouragement. Assistance also was provided by Joanne Girvin of Gallatin National Forest, Montana, Ann Poore of *The Salt Lake Tribune,* and Sandra Rush, representing the Geological Society of America. We also thank Jack Shea and the staff at the Teton Science School and Whit Clayton, Bob Dornan, Dick Barker, and other folks from Moose, Wyoming, who have followed our work and provided encouragement.

Financial support for the long-term research of Bob Smith in the Yellowstone–Teton area has been provided by the National Science Foundation, the U.S. Geological Survey, the National Park Service, and the University of Utah's Department of Geology and Geophysics. We gratefully acknowledge the support for this publication by the Grand Teton Historical Association and Sharlene Milligan.

Bob Smith greatly appreciates the care and interest in this book from his wife Janet; children Elissa Richards, Rob Smith, and Nicole Smith; and mother Gretta Smith. His family accompanied him on many research trips in Yellowstone and Grand Teton national parks and enthusiastically supported his work.

Lee Siegel thanks his friends for their encouragement, and is indebted to James E. Shelledy, Terry Orme, Shia Kapos, and Tim Fitzpatrick at *The Salt Lake Tribune* for permitting his absences to work on this book. He also appreciates his former teachers in Oregon—William H Oberteuffer and the late David C. Park and Alice Metcalf—for inspiring a lifelong love of science and learning. Siegel is particularly grateful for the loving support of his family, particularly for the encouragement and kindness of his father Sol, who died July 6, 1999.

Windows into the Earth

A Land of Scenery and Violence

1

Anyone who has spent summers with pack-train in a place like Yellowstone comes to know the land to be leaping. . . . The mountains are falling all the time and by millions of tons. Something underground is shoving them up.

—*T. A. Jaggar, Hawaiian Volcano Observatory, 1922*

It was the busy summer season in Yellowstone National Park, a beautiful moonlit night with 18,000 people in the park's campgrounds and hotels and thousands more in surrounding towns and recreation areas. At 23 minutes before midnight, a talent contest was wrapping up at the Old Faithful recreation hall. A beauty queen had just been crowned. As she walked down the aisle to the applause of several hundred people, the log building creaked loudly and began to shake. Within seconds, the earthquake sent people scurrying for the exits. A park ranger dropped the hand of his date—a waitress from Old Faithful Inn—and rushed to open the doors so no one would be trampled. Nearby, frightened guests fled Old Faithful Inn, where a waterline broke and an old stone chimney soon would collapse into a dining room, thankfully closed at that late hour.

Out in the darkness, in geyser basins along Yellowstone's Firehole River, the Earth began belching larger-than-usual volumes of hot water. About 160 geysers erupted, some for the first time, others after decades-long dormant periods. Sapphire Pool, once a gentle spring, became a violent geyser, hurling mineral deposits around Biscuit Basin. Clepsydra, Fountain, and some other geysers in Lower Geyser Basin began erupting more often than usual. Old Faithful's eruptions became less frequent, although some observers thought it spouted with unusual vigor earlier that evening. Hundreds of hot springs became muddy. Fountain Paint Pot spewed mud violently, spattering tourist walkways.

Rocks and landslides tumbled into park highways in several places, blocking roads between Old Faithful and Mammoth and closing the route to the park's west entrance at West Yellowstone, Montana. Within an hour, thousands of vehicles streamed out of Yellowstone on roads that remained open—a serpentine parade of headlights fleeing the strongest earthquake yet recorded in the Rocky Mountains and the Intermountain West.

The panic and damage in Yellowstone were minimal compared with the unimaginable horror that would overtake a popular Montana recreation area just outside the park's northwest boundary.

Along the shores of 15-mile-long Hebgen Lake, 10 miles northwest of West Yellowstone, residents and vacationers were sleeping. The lake was impounded behind the 87-foot-tall Hebgen Dam, built in 1915. Below the dam was the Madison River Canyon, one of America's best trout fishing areas. A few hundred people were camped in the 9-mile-long canyon, either in turnouts along the highway or in campgrounds such as Rock Creek, about 7 miles below the dam.

Melvin Frederick of Elyria, Ohio, was sleeping at Rock Creek in a tent with his son and nephew, both 15. His wife and 16-year-old daughter slept in the family's 1958 Plymouth station wagon. The shaking started suddenly. Frederick recalled: "At first, I thought it was a bear tearing through the nearby trees. Someone shouted, 'It's a tornado or an earthquake!' There was a tremendous roar. Outside the tent and looking upward, I saw the whole side of a mountain collapse. It looked like a huge waterfall. There was a gush of air, followed by a wave of water from the Madison River."

The wave hit, sweeping Frederick's son, Paul, 50 feet and pinning the teenager between trees and a trailer. Water rose to Paul's chin, and it looked like he would drown, but the rising water finally floated the trailer upward, allowing Paul's father and cousin to pull him free. Frederick's wife and daughter screamed as their station wagon rolled over in the water. They got out just as the car rolled again. Paul and his cousin were badly hurt, but the family managed to reunite and climb to higher terrain. There, they knelt on the ground and prayed.

Prayers could not help others at Rock Creek. Truck driver Purley Bennett, 43, and his wife, Irene, 39, were sleeping in a trailer. Camped outside were the Coeur d'Alene, Idaho, couple's four children: Phillip, 16; Carole, 17; Tom, 11; and Susan, 5. Soon after the quake, "There was a huge roar," Phillip said later. "I looked up and saw the mountain cascading down upon us."

His parents, stepping out of their trailer, were hit by wind so strong it ripped off their clothing. Purley Bennett grabbed a tree. Then the wave of water hit the family, blasting them away. Irene Bennett briefly lost consciousness, awakening to find herself naked on the river bank and pinned beneath a downed tree. She dug herself out, and called for her family all night. Phillip, in pain with a broken leg, had managed to crawl to safety, digging a hole and covering himself with dirt for warmth. After daylight, he crawled to his mother.

Lloyd Verlanic of Anaconda, Montana, was the first to reach Irene Bennett and her son. "She was crawling on her hands and knees," Verlanic recalled. "She looked up at me, her face a mass of blood, and said, 'Oh, thank God! I've been crawling and praying all night. My children, oh my God, my children. I wish I could find my children.' " Searchers soon found her husband's body.

A day later, recovering in an Ennis, Montana, hospital, Irene Bennett told a radio reporter she had been informed her other children were dead, "but I have not lost faith in God. I know they will be found." Out in the hallway, a nurse said searchers that morning recovered the bodies of the three children. Only Irene and Phillip had survived. "The poor woman; she just does not realize yet," the nurse said.

In the terror and confusion, it was difficult for most people in Madison River Canyon to realize what had happened.

At 11:37 P.M. Mountain Time on Monday, August 17, 1959, a magnitude-6.3 foreshock was followed within seconds by a major earthquake of magnitude 7.5. The strong shaking lasted less than a minute. It sent an entire mountainside crashing down at 100 miles per hour on the lower end of the Rock Creek campground. The slide flowed from south to north, falling more than 1,000 feet and plunging at a right angle into Madison River Canyon, which ran from east to west. As the slide roared downhill, it pushed winds of nearly hurricane force in front of it. Then the landslide slammed into the Madison River, slapping the river from its bed and hurling 30-foot waves upstream and downstream. The waves tossed cars, trailers, and people like toys. The slide filled the canyon bottom and raced more than 400 feet up the north side of the canyon, dumping boulders as big as houses. The slide not only buried part of the campground, but covered a mile-long stretch of highway and river with 100 to 300 feet of boulders, rocks, and soil. After the slide halted and the waves subsided, the Madison River started backing up behind the slide, forming a new lake that began inundating what was left of the campground (Figure 1.1).

It took weeks for authorities to determine the toll: twenty-eight dead, dozens injured, and $13 million in damage. Of the twenty-eight fatalities, nineteen were entombed forever beneath the giant Madison landslide. Five bodies were recovered in

1.1 ~ *The giant Madison Canyon landslide was triggered by the Hebgen Lake earthquake. Eighty million tons of rock plunged off the south side of Madison Canyon, 15 miles northwest of West Yellowstone, Montana. Many of the quake's victims were buried by the slide, which dammed the Madison River to create Earthquake Lake. (U. S. Geological Survey.)*

the slide area. Some had drowned. Two people, severely injured by the slide, died later in a Bozeman, Montana, hospital. The other two victims died about 7 miles southwest of the slide, crushed by a boulder while they slept.

The earthquake generated large waves in nearby Hebgen Lake and caused its north shore to drop 19 feet, dumping a number of homes and several stretches of highway into the lake. Buildings were damaged in West Yellowstone and more distant Montana towns. The quake was felt through much of the West, including Wyoming, Montana, Idaho,

California, Nevada, Utah, Oregon, Washington, and British Columbia. Seismograph needles were knocked off scale hundreds of miles away in Salt Lake City and Denver. The seismic waves made water levels fluctuate in wells thousands of miles away in Hawaii, New Jersey, Florida, and Puerto Rico.

A Hotspot for Disaster

The Hebgen Lake earthquake happened because huge blocks of rock, as much as 15 miles long, abruptly slumped and tilted beneath the lake and Madison River Canyon. The ultimate reason these blocks moved was that Earth's crust is being stretched apart in a broad region of the West known as the Basin and Range Province. Such stretching has created valleys, mountain ranges, and earthquake faults from eastern California through Nevada and western Utah, and north into Oregon, Idaho, Montana, and western Wyoming.

However, the Hebgen Lake quake was unusual. Not only was it the strongest ever recorded in the Intermountain West and the Rockies, but it produced extraordinary ground movements. Two parallel segments of the Hebgen Lake fault broke the ground surface north of Hebgen Lake and Madison Canyon. They were "normal" faults. Quakes on normal faults occur when ground on one side of the fault drops down and away from ground on the other side, which rises upward. Such vertical movement creates small cliffs or embankments called scarps. In some places along the faults broken during the Hebgen Lake quake, these abrupt cliffs were as high as 22 feet (Figure 1.2). So much vertical "offset" or "displacement" was unheard of for a normal fault earthquake in the West. In prehistoric times, such quakes typically produced 6 to 10 feet of vertical movement.

The quake's large ground movements suggest the disaster was not caused solely by stretching of Earth's crust in the West. Another great geological force may have contributed to the quake's power—a force recognized by volcanologist T. A. Jaggar decades before modern knowledge about earthquakes. Jaggar was right. Something underground is shoving Yellowstone upward and helping make the land leap and the mountains fall during earthquakes.

That *something* is one of the Earth's great unseen geological features: the Yellowstone hotspot—a huge, vertical plume of hot rock that extends upward from at least 125 miles beneath Yellowstone National Park and includes zones where rock is molten or partly molten. Sitting atop the hotspot, the Yellowstone region is essentially a giant, slumbering volcano that huffs upward and puffs downward over the

1.2 ~ The Hebgen Lake fault broke vertically during the magnitude-7.5 Hebgen Lake, Montana, earthquake. Ground on the right side of the fault dropped as much as 22 feet relative to land on the left. (Robert B. Smith.)

decades, like a breathing beast. The Hebgen Lake quake may have been powered in part by the heat and molten rock within the hotspot lifting the region upward, thus hastening the stretching of Earth's crust in the area.

If so, the Hebgen Lake earthquake would be only the latest in a long string of catastrophes triggered by the violent geology of the Yellowstone hotspot. During the past 16.5 million years, geological processes related to the hotspot have included gigantic volcanic eruptions, mountain building, major earthquakes, geothermal activity, landslides, and even glaciers, floods, and forest fires. These disasters sculpted Yellowstone's fantastic, high-altitude scenery and reshaped the landscape of about a quarter of the northwestern United States, including parts of Idaho, Oregon, Nevada, Utah, Wyoming, and Montana.

During roughly the same time, the stretching of Earth's crust in the Basin and Range Province shaped an even larger region of the West, including Grand Teton National

Park, just south of Yellowstone. Crustal stretching repeatedly caused major earthquakes on the Teton fault during at least the past 13 million years, helping to build the spectacular Teton Range.

The combination of geologic processes at work in the Yellowstone–Teton region is seen nowhere else on Earth on such a large scale and with such vivid manifestations. In a sense, Yellowstone and Teton geology is a living entity. It belches violently during volcanic eruptions, blowing away or swallowing whole mountain ranges, repaving valleys, dumping ash on half a continent, and changing global climate. It fractures the ground and uplifts mountains with earthquakes. And it emits enormous thermal energy to produce the world's greatest concentration of geysers and hot springs.

The multiple disasters that molded the landscape and environment of the Yellowstone–Teton region reveal secrets of dynamic geological processes hidden within the Earth: the volcanic power of the Yellowstone hotspot and the crustal movements ripping apart the West. In that sense, Yellowstone and Grand Teton national parks—with their scenic glories and natural disasters—serve as windows that let us peer into Earth's interior.

Yellowstone Hotspot Volcanoes Shaped the West

Hotspots help shape Earth's surface as they release heat from the planet's interior through volcanic eruptions and hydrothermal activity, which is the activity of hot water in geysers, hot springs, and steam vents. The Yellowstone hotspot is the largest hotspot under a continent and among the largest of some thirty active hotspots on Earth. Most hotspots are beneath the world's ocean bottoms. Giant plates of Earth's crust make up the seafloor, and these plates slowly drift over erupting hotspots, creating island chains such as Hawaii. Other hotspots lie beneath Iceland, the Galápagos Islands, and the Azores.

The North American plate of Earth's crust has drifted southwest over the Yellowstone hotspot at about 1 inch per year. About 16.5 million years ago, the hotspot was beneath what is now northern Nevada near its border with Oregon and Idaho. As the continent drifted southwest, the hotspot in effect moved northeast beneath Oregon's southeast corner, then across Idaho to Wyoming. It ended up beneath Yellowstone about 2 million years ago.

The hotspot's passage was far from peaceful. Roughly 100 times during the past 16.5 million years, huge volumes of magma, or molten rock, rose from the hotspot and erupted catastrophically through a giant crater—known as a caldera—measur-

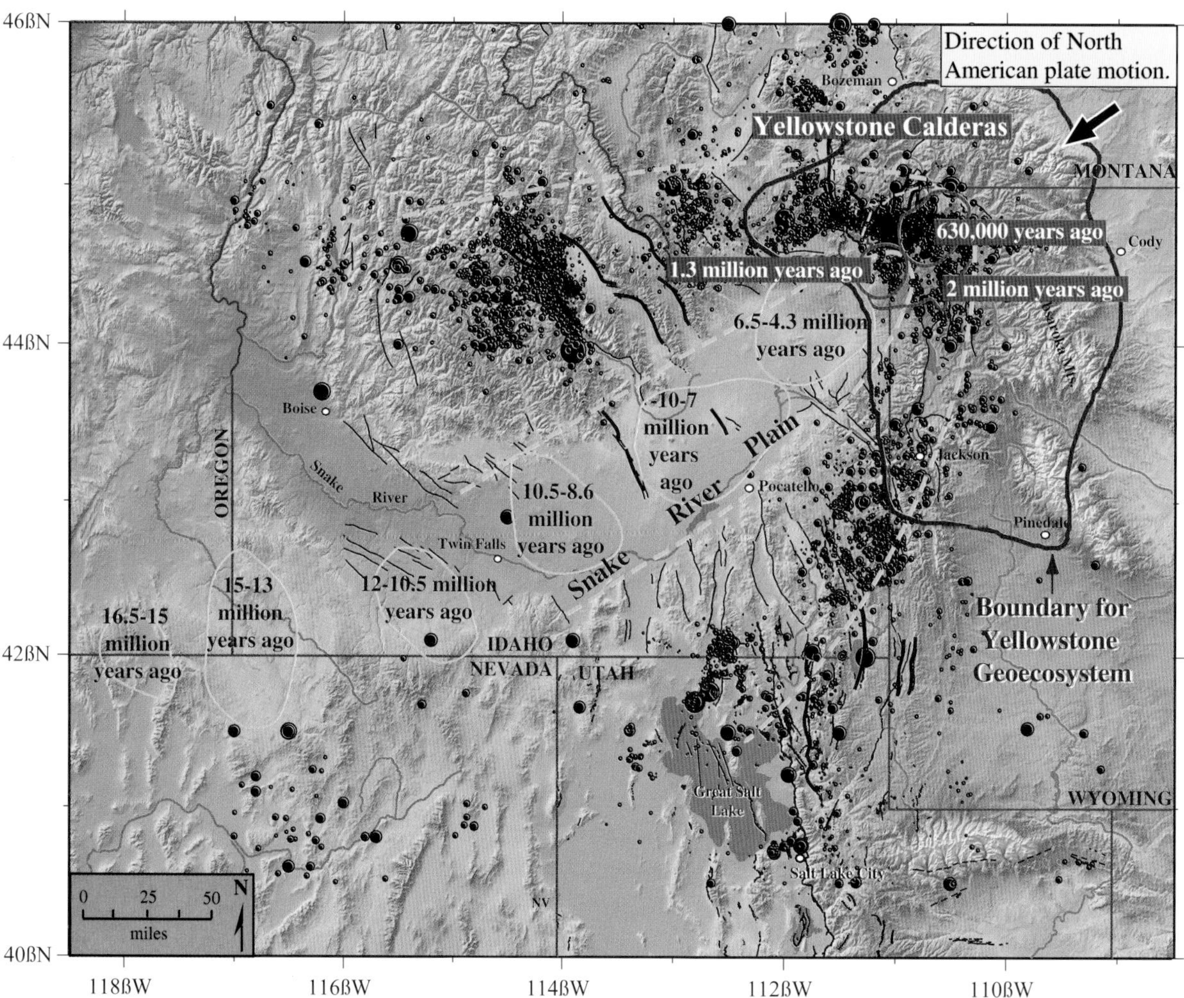

1.3 ⁓ *Path of the Yellowstone hotspot. Yellow and orange ovals show volcanic centers where the hotspot produced one or more caldera eruptions—essentially "ancient Yellowstones"—during the time periods indicated. As North America drifted southwest over the hotspot, the volcanism progressed northeast, beginning in northern Nevada and southeast Oregon 16.5 million years ago and reaching Yellowstone National Park 2 million years ago. A bow-wave or parabola-shaped zone of mountains (browns and tans) and earthquakes (red dots) surrounds the low elevations (greens) of the seismically quiet Snake River Plain. The greater Yellowstone "geoecosystem" is outlined in blue. Faults are black lines.*

ing as much as a few tens of miles wide. Because North America was drifting southwest over the hotspot, the eruptions formed a chain of progressively younger calderas stretching from the Idaho–Oregon–Nevada border 500 miles northeast up Idaho's Snake River Plain to Yellowstone in Wyoming (Figure 1.3). Many of these calderas overlapped, so the approximately 100 caldera eruptions occurred in at least seven and possibly thirteen "volcanic centers."

Giant caldera eruptions happen so infrequently that none have occurred in recorded history, but they are the largest volcanic eruptions on Earth. As caldera-forming eruptions progressed across Idaho to Yellowstone during the past 16.5 million years, they repeatedly dumped volcanic ash on half the United States while existing mountains either were blown away or sank into the giant craters.

When the hotspot passed under an area, it created a 300-mile-wide, 1,700-foot-tall bulge in Earth's crust, like a rug being dragged over an object. So the old calderas—essentially "Old Yellowstones"—now hidden beneath the Snake River Plain once were at higher elevations. However, in the wake of the hotspot, the plain sank as much as 2,000 feet and was flooded by a series of basalt lava flows, which covered the old calderas. What was left was a vast valley—the eastern Snake River Plain. Idaho's famous potatoes grow in its fertile volcanic soils.

Yellowstone drifted over the hotspot sometime before 2 million years ago. Molten rock rising from the hotspot produced three cataclysmic caldera eruptions more powerful than any in the world's recorded history. First, the Yellowstone caldera exploded 2 million years ago. Then, the activity shifted several miles west into Idaho, where the Island Park caldera—also called the Henry's Fork caldera—blew up 1.3 million years ago. Finally, activity moved east again and another eruption 630,000 years ago formed the modern Yellowstone caldera, which overlaps remnants of the 2-million-year-old caldera. The three eruptions, respectively, were 2,500, 280, and 1,000 times larger than the 1980 eruption of Mount St. Helens in Washington state.

The Yellowstone and Island Park calderas are considered a single volcanic center. Together, their three catastrophic eruptions expelled enough ash and lava to fill the Grand Canyon. They belched volcanic debris high into the atmosphere and covered much of the West and Great Plains with volcanic ash. Aerosols and gases from the eruptions likely cooled Earth's climate for years, producing so-called "volcanic winters."

Visitors expecting rugged Rocky Mountains often are surprised by the undulating, relatively flat, forested topography of the Yellowstone Plateau, much of which is above 7,700 feet (Figure 1.4). The landscape is a legacy of the violent caldera eruptions. While the eruptions may have blown away some of the ancient mountains that once crossed Yellowstone, it is more likely the mountains were swallowed by the

1.4 ⁓ Space view of Grand Teton and Yellowstone national parks from satellite images overlaid on digital elevation maps. The 8,000-foot-high Yellowstone caldera (marked III) was produced by a giant volcanic eruption 630,000 years ago. The caldera occupies a 45-by-30-mile-wide area of central Yellowstone. The Teton fault bounds the east side of the Teton Range and raised the mountains high above Jackson Hole's valley floor. Partial boundaries of the calderas formed 2 and 1.3 million years ago are marked I and II, respectively. (Computer image by E. V. Wingert.)

Earth as the caldera floor sank downward during the explosive blasts. The central portion of Yellowstone National Park—including the west half of Yellowstone Lake—sits inside the 45-by-30-mile-wide caldera formed by the eruption 630,000 years ago. The crater rim is evident in some places, such as north of Madison Junction. However, almost thirty subsequent smaller lava eruptions, most recently 70,000 years ago, filled in the caldera and surrounding portions of the Yellowstone Plateau.

The landscape also was smoothed by giant prehistoric glaciers and sediments deposited in ancient lakes.

With all these changes obscuring the complete outline of the Yellowstone caldera, it is somewhat surprising the giant crater was accurately recognized by Ferdinand V. Hayden, whose early explorations prompted Congress to make Yellowstone the first U.S. national park in 1872. A year earlier, Hayden stood atop Mount Washburn, a 10,243-foot peak just north of the caldera, and recognized he was looking into a giant volcanic system:

> This basin has been called by some travelers the vast crater of an ancient volcano. It is probable that during the Pliocene period the entire [Yellowstone] country . . . was the scene of as great volcanic activity as that of any portion of the globe. It might be called one vast crater, made up of thousands of smaller volcanic vents and fissures out of which the fluid interior of the earth, fragments of rocks, and volcanic dust were poured in unlimited quantities.

Today, the ground at Yellowstone emits thirty to forty times more heat than the average for North America. That heat powers Yellowstone's hydrothermal attractions—the world's greatest display of geysers, hot springs, mud pots, and steam vents. The heat comes from the hotspot.

For years, the popular notion of a hotspot was of a continuous plume of molten rock extending from deep within the Earth, perhaps even its molten core. Research indicates this is true of most hotspots, but the Yellowstone hotspot may be more complex. The best available evidence suggests the Yellowstone hotspot likely begins in Earth's upper mantle, about 125 miles beneath the surface. There, rising heat from deeper within the Earth melts rock. The hotspot has a plume shape, but the plume is not completely molten. Rather, it is a column of hot rock in which some molten rock flows upward. The hotspot plume extends upward until it hits the colder, overlying North American plate of Earth's crust and uppermost mantle. There, at a depth of about 50 miles, blobs of molten rock—iron-rich basaltic magma—periodically rise through the rest of the upper mantle and into Earth's crust. Heat from these blobs of molten rock melts overlying, silica-rich rock to create a large, sponge-like chamber of partly molten rhyolite. The magma chamber begins about 5 miles beneath the Yellowstone caldera and extends down to 8 miles deep, but may be as shallow as 3 miles in the northeast caldera.

This magma chamber heats overlying salty water, which in turn heats shallower groundwater to produce more than 1,000 geysers, hot springs (Figure 1.5), and mud

1.5 ~ *Aerial view of Grand Prismatic Spring in Midway Geyser Basin. This 200-foot-wide, vividly colored hot spring exemplifies Yellowstone's world-renowned geothermal features. Mineral deposits next to the spring are colored by microbes that thrive in hot water.*

pots—Yellowstone's main attractions. These hydrothermal features can become clogged by mineral deposits within their plumbing systems, but earthquakes periodically shake the conduits to break up the clogging minerals. Many small earthquakes are triggered by movement of molten rock and superheated water beneath Yellowstone. So the geysers owe their survival to ongoing earthquake activity that ultimately is driven by the hotspot.

Magma and hot water beneath Yellowstone also make the floor of the caldera huff upward and puff downward over the years. Using surveying instruments to measure elevation changes, scientists determined much of central Yellowstone rose by more than 3 feet from 1923 to 1984, then started sinking.

As the caldera floor dropped, thousands of small quakes rattled Yellowstone during 1985 and 1986. The pattern of epicenters from the quakes suggests hot water and perhaps molten rock drained from beneath the caldera, moving northwest through faults that line up with the Hebgen Lake fault zone farther northwest. By 1995, the caldera floor had dropped 8 inches. Then, in 1995 and 1996, it started rising again. Such "breathing" is typical of calderas, which can huff and puff for tens or hundreds of millennia without erupting.

The subterranean movements of hot water and molten rock only occasionally burp lava onto the surface and rarely explode in a violent caldera-forming eruption. It is impossible to predict when more eruptions will strike Yellowstone. Another catastrophic caldera eruption appears likely within tens of thousands of years to hundreds of thousands of years and thus is unlikely in our lifetimes. Smaller but significant eruptions, like the 1980 explosion of Mount St. Helens, may occur within thousands of years. Small steam and hot-water explosions would not be surprising within centuries or even decades.

Burn, Freeze, Shake, and Slide

Volcanism is far from the only disaster produced by the hotspot. Yellowstone already sat high in the Rocky Mountains but was pushed to loftier heights—above 7,700 feet—atop the hotspot's broad, upward bulge. The high elevation and resulting climate help determine the plants and wildlife that thrive in Yellowstone. For example, the altitude and Yellowstone's volcanic soils foster vast forests of lodgepole pine, which undergo a natural 300-year cycle of growth, devastation by fire, and rebirth. In 1988, not quite three decades after the big quake, a major inferno burned more than a third of Yellowstone National Park.

The lofty heights also helped a 3,500-foot-thick icecap form atop the Yellowstone Plateau during at least three global glacial episodes within the past 250,000 years to 2 million years. The Yellowstone ice field was so large it covered most of Yellowstone and Grand Teton national parks—an area extending more than 100 miles north–south and 70 miles east–west. Large glaciers flowed downhill from the plateau and other mountains. After volcanism shaped Yellowstone's landscape and the Teton fault produced the terrain of the Teton Range and Jackson Hole, Wyoming, the Ice Age glaciers left their own marks. They sharpened the spires of mountains. They helped carve valleys such as those occupied by the Snake and Yellowstone rivers. The glaciers also excavated smaller lakes at the base of the Teton Range and deepened Yellowstone and Jackson lakes. Each time the glaciers receded—the last time was about 14,000 years ago—their meltwaters produced mighty floods that accentuated the downcutting of valleys, including the breathtaking Grand Canyon of the Yellowstone River.

Of all the geological processes fostered by the hotspot, earthquakes are the most dominant on a human timescale. Although earthquakes happen throughout the Basin and Range region, they are more frequent above and around the Yellowstone hotspot. During geological time, major earthquakes rocked the Yellowstone–Teton region repeatedly. Today in the conterminous United States, only faults in California produce more earthquakes than the broad area encompassing Yellowstone National Park. With the hotspot beneath Yellowstone, and the West still stretching apart, more strong earthquakes are likely in the future—both near Yellowstone and along the Teton fault to the south.

The hotspot influences the pattern of seismic activity through a broad area of Utah, Wyoming, Idaho, and Montana. When a boat moves through water, it creates an arc- or parabola-shaped bow wave extending from the front of the boat backward along both sides. In a similar manner, as the Yellowstone hotspot moved across Idaho to Yellowstone, it left in its wake an arc-shaped belt of mountains and earthquake epicenters bordering both sides of the Snake River Plain lowlands. (The epicenter of an earthquake is the location on Earth's surface directly above the part of a fault where the quake began to rupture the fault.) Few earthquakes occur on the plain itself. The mountains flanking the Snake River Plain are made of cold, brittle crust conducive to rupturing during earthquakes.

The wake of earthquake epicenters flanking the Snake River Plain includes the region's two strongest historic earthquakes: the 1959 Hebgen Lake quake and the 1983 magnitude-7.3 Borah Peak, Idaho, quake, which killed two people in Challis, Idaho. It also includes the Teton fault, one of several faults in or near the Yellowstone–Teton region that could generate major quakes in the future. Although an ancestral Teton fault may have existed as long as 34 million years ago, the fault became active in its present form about 13 million years ago, and perhaps more recently. Since then, perhaps a few thousand major earthquakes lifted the Teton Range into its towering setting while simultaneously making the valley of Jackson Hole sink.

No major quake has happened on the Teton fault in historic time. Yet smaller quakes have spelled disaster. Decades before the Hebgen Lake quake and the Madison Canyon landslide, a much bigger landslide likely was triggered by a small quake in or near Jackson Hole. One night in 1925, the valley was rattled by a jolt of about magnitude 4. (Magnitude is a measure of energy released by an earthquake.) The next day, a waterlogged mountainside collapsed into Gros Ventre Canyon on the east edge of Grand Teton National Park. The slide was one-third larger than the Madison slide. No one was killed by the slide itself. However, the slide dammed the Gros Ventre River to form a lake. Two years later, in 1927, the natural dam burst, flooding downstream towns and killing six people.

Today, the Teton fault and Utah's Wasatch fault are the most widely recognized seismic hazards in the Intermountain Seismic Belt, a band of quake activity extending from Montana through Idaho, Wyoming, and Utah to southern Nevada. Most Utah residents live atop the Wasatch fault, many in vulnerable unreinforced brick homes. The Teton fault threatens eventual disaster to thousands of people visiting Grand Teton National Park and the resort town of Jackson, Wyoming. The Wasatch and Teton faults are believed capable of major quakes of about magnitude 7.5, and both faults may be due or overdue for such a disaster.

Although humans cannot prevent earthquakes, volcanic eruptions, most landslides, or other natural disasters, their death toll and destruction can be reduced by judicious land use, monitoring seismic and volcanic activity, alerting the public to these potential catastrophes, and planning to cope with the disaster. In planning for major quakes on the Teton, Wasatch, and other faults in the region, emergency management officials use the 1959 Hebgen Lake earthquake as the worst-case or "maximum credible earthquake" possible in the region, both because of its large magnitude and the extraordinarily large fault movements it produced. Of the disasters that have struck the Yellowstone–Teton region in historic time, none remind us of the immense power and unpredictability of Earth's geologic forces more than the Hebgen Lake quake and Madison Canyon landslide.

A Night of Terror

The size of the Madison slide is difficult to fathom. Its volume totaled 37 million cubic yards and it weighed 80 million tons—enough to pave a two-lane highway 3 feet deep from Montana to New York City. No statistics, however, can capture the terror felt by those trapped in Madison River Canyon that August night in 1959.

Mrs. Clarence Scott, 57, of Fresno, California, was sleeping with her husband in their trailer at Rock Creek campground when the landslide hurled the Madison River from its bed. The waves hit the trailer, ripping it open and throwing out Mrs. Scott. "It was horrible," she said. "Children were screaming and crying for their mothers. And husbands were begging for their wives to answer."

Henry Bennett and his family had traveled from Arizona to Rock Creek, where they befriended Sydney and Margaret Ballard and their disabled son, Christopher.

Bennett said later: "When the quake hit, we heard them scream and watched as they tried to get out, but water swept them away. They're surely dead."

Grover Mault and his 68-year-old wife had been camped at Rock Creek for a week. The California couple awoke when their trailer was knocked end-over-end, then hurled into the water. Mault and his wife, clad in nightclothes, managed to get out and climb atop the trailer as it floated in ever-deepening water. They grabbed tree limbs as the trailer sank away. The limbs kept breaking as the Maults clutched at higher branches. They screamed for help. Mault's wife went under repeatedly. "She wanted me to let her go, but I told her that if she went, I'd go too," Mault recalled. As he kept fishing his wife out, he saw more landslides tumble down the mountainside. "I thought the world had come to an end." The Maults stayed in the tree all night as the water rose. Tuesday morning, they were rescued.

Some 150 survivors, fearing Hebgen Dam would collapse, climbed to a flat area above the river now known as Refuge Point. By 2 A.M. Tuesday, less then 3 hours after the quake, some of them formed a rescue party, went to the landslide, and returned with twenty people, several with serious injuries.

As death and destruction crashed on Madison Canyon, it also touched a campground at Cliff Lake, roughly 7 miles southwest of the landslide. The Stryker family of San Mateo, California, had visited Cliff Lake for years. When the quake hit, the three Stryker boys—Martin, 15; John, 13; and Morgan, 8—were asleep in their tent. Their parents slept in another tent nearby. "We could hear trees cracking," one of the boys said. "When we got out of the tent, two trees had fallen across our car and another crushed our boat." Their parents were dead, crushed by a large boulder. "They couldn't have known what hit them," Martin said.

In the mountains above Madison Canyon and Hebgen Lake, the earthquake cracked the ground along two long segments of the Hebgen Lake fault. Along each segment, ground on one side of the fault rose up while ground on the other side dropped down, creating a cliff-like embankment, or scarp, as high as 22 feet. One crack, known as the Hebgen fault scarp, ran for 8 miles along the north shore of Hebgen Lake and down into Madison Canyon. The Hebgen scarp ran smack through the Cabin Creek campground below Hebgen Dam. Vehicles were trapped when the exit road was cut off by the scarp. The other fault scarp, known as the Red Canyon segment, started near the east end of Hebgen Lake and stretched north and then west through the mountains for about 14 miles.

It was not the fault scarps, but the tilting of huge blocks of crustal rock along those faults that caused destruction and panic along the shores of Hebgen Lake. The ground beneath the lake dropped downward along the Hebgen Lake fault and also tilted northward toward the fault lines and the mountains. The tilting made ground on the lake's north shore drop as much as 19 feet, permanently inundating homes, cabins, docks, and other facilities. Meanwhile, the lake's south shore tilted upward several feet, leaving docks and boats high and dry.

Shaking from the quake triggered relatively small landslides that dumped sections of Highway 287 into the lake. Madison Canyon was cut off by the highway damage upstream and the slide downstream. Along Hebgen Lake, shaking damaged many homes, knocking down chimneys and throwing buildings off their foundations. At Hilgard Lodge, a mile above Hebgen Dam, resort owner Grace Miller was shocked awake as her cabin started sliding into the lake as large fissures opened in the slumping ground. She made it to safety with her dog.

The shaking and tilting of the ground made water in the lake slosh back and forth in what is called a seiche—the lake equivalent of a quake-triggered tsunami or "tidal wave" in the ocean. Every 17 minutes, the water washed high on one shore, withdrew, then surged again. The initial waves were 20 feet high. At Hebgen Dam, caretakers George Hungerford and Lester Caraway feared the quake might have weakened the 44-year-old earth-filled, concrete-cored structure. From a vantage point below the dam, they saw a 4-foot-tall wave come toward them and fled for high ground, fearing the dam was collapsing. Instead, the seiche wave reached the dam, then rose 3 or 4 feet above the top. Hungerford and Caraway watched the wave pour over the dam for 5 to 10 minutes, then recede. The seiche sloshed back and forth for more than 11 hours, but only the first several sloshes were tall enough to send water over the dam. The dam's concrete core and spillway were cracked, and some fill settled several feet, but the dam held.

Six hundred residents of Ennis, Montana, more than 40 miles below the Madison landslide, awoke to the sound of a wailing siren and sought high ground in the darkness, waiting for a flood that never came.

Behind the dam, the seiche waves smashed through lakeshore cabins and homes already ravaged by the shaking, then inundated by lake water when the ground dropped. Some trailers were carried into the lake, their occupants narrowly escaping. In homes on higher ground, chimneys, stoves, and refrigerators toppled. Large fissures opened in the ground.

Near the east end of the lake, a panicked guest fled the Duck Creek Cabins. He sped away in a Cadillac, but in the darkness failed to see that the fault had ripped the highway apart, creating a 12-foot-high embankment. The car plunged over it and flipped upside-down. The man smashed a window to escape.

Ten miles south, in West Yellowstone, Los Angeles resident Laura Schauer was walking out of the Wagon Wheel Cafe when suddenly, "I felt like I was on waves. I had no control. The lights swung. Glass was flying everywhere. I picked myself up off the ground several feet from the steps." Elsewhere in town, chimneys toppled. Large windows shattered. Goods were thrown from store shelves. Chimney stones collapsed at the Union Pacific depot, and 250 windows broke. Phone service and electricity were knocked out. Leonard Kelly, who ran the Hitchin' Post Motel, was bashed back and forth in his doorway as he tried to run outside, dishes crashing and furniture jumping behind him. Within hours, most of his guests were gone. "They didn't check out. They just left," Kelly's wife said.

In and near Yellowstone, the quake led many people to believe bears were attacking. Some stonework collapsed at Mammoth Hot Springs, the park headquarters. Old Faithful Inn was shut down. Yellowstone's west entrance was closed.

The quake broke windows and toppled chimney bricks in Butte, Dillon, and Virginia City, Montana. In Deer Lodge, the Montana Prison's original cellblock was damaged so badly it was evacuated and later demolished. In Billings, Montana, downtown buildings swayed and residents were jarred awake. In a few homes, plaster cracked. One tavern patron declared, "That's the last time I'll ever touch the stuff." Another demanded a refill.

"There Is Nothing More to Find"

Soon after daylight on Tuesday, planes and helicopters were dispatched to the disaster zone from bases in Utah, Idaho, California, Nevada, and Montana. Forest Service smokejumpers parachuted to Refuge Point in Madison Canyon, preparing the injured for evacuation. Helicopters later swooped down, picked up the injured, and ferried them to West Yellowstone, which was nearly a ghost town. From there, the seriously injured were flown to hospitals in Bozeman, where two died.

Rescue workers and friends and relatives of the missing searched Madison Canyon. Bulldozers plowed roads to allow stranded campers to leave. Divers plunged into the new lake forming behind the slide. By late Tuesday, less than 24 hours after the quake, Madison County Sheriff Don Brooks said: "We airlifted between 100 and 150 stranded campers and brought five bodies out of the area."

The bodies, found downstream from the slide, were those of Purley Bennett, three of his children, and Thomas Mark Stowe from Sandy, Utah. Searchers never found nineteen other victims, including Stowe's wife. Her father came to the disaster scene and dug into the slide until his hands bled.

By the weekend, the search ended, although most of the dead—the number wasn't known at the time—remained buried under the massive slide. Montana Civil Defense Director Hugh Potter said: "There is nothing more to find."

Aftershocks rattled the area daily. The 7.5 quake was followed within 10 hours by five aftershocks between 5.5 and 6.3 in magnitude. By Friday morning, the fourth day after the quake, 370 aftershocks had been recorded. Five years later, on October 21, 1964, the area was slapped by a magnitude-5.8 aftershock.

Army Corps of Engineers officials began surveying the landslide to determine how to prevent the rising waters of newborn Earthquake Lake from cutting through the natural dam and flooding the Madison River Valley and the town of Ennis below. Nine days after the quake, the Corps began bulldozing a spillway through the landslide. Two weeks after the quake, the lake was 5 miles long and 150 feet deep, rising as much as 9 feet daily. The Corps worked feverishly to excavate the mile-long spillway. They finished in time. Late September 9, the waters of Earthquake Lake reached the top of the slide and began flowing safely over the spillway.

The American Red Cross took thousands of calls from people who feared that relatives were among the dead. Most turned up safely. Some never did.

Roger Provost, his wife, and their two sons were among those buried beneath the slide. Their deaths were not confirmed until three weeks after the earthquake, when Provost failed to return to work in California. The last word from the family was a postcard dated August 16, the day before the quake. Provost had written to his mother that he and his family were camped along the Madison River.

In the Wake of the Yellowstone Hotspot

2

Anyone who drives through southern Idaho on Interstates 84 or 15 must endure hours and hundreds of miles of monotonous scenery: the vast, flat landscape of the Snake River Plain. In many areas, sagebrush and solidified basalt lava flows extend toward distant mountain ranges, while in other places, farmers have cultivated large expanses of volcanic soil to grow Idaho's famous potatoes.

Southern Idaho's topography was not always so dull. Mountain ranges once ran through the region. Thanks to the Yellowstone hotspot, however, the pre-existing scenery was destroyed by several dozen of the largest kind of volcanic eruption on Earth—eruptions that formed gigantic craters, known as calderas, measuring a few tens of miles wide.

Some 16.5 million years ago, the hotspot was beneath the area where Oregon, Nevada, and Idaho meet. It produced its first big caldera-forming eruptions there. As the North American plate of Earth's surface drifted southwest over the hotspot, about 100 giant eruptions punched through the drifting plate, forming a chain of giant calderas stretching almost 500 miles from the Oregon–Nevada–Idaho border, northeast across Idaho to Yellowstone National Park in northwest Wyoming. Yellowstone has been perched atop the hotspot for the past 2 million years, and a 45-by-30-mile-wide caldera now forms the heart of the national park.

After the ancient landscape of southern and eastern Idaho was obliterated by the eruptions, the swath of calderas in the hotspot's wake formed the eastern two-thirds of the vast, 50-mile-wide valley now known as the Snake River Plain. The calderas eventually were buried by basalt lava flows and sediments from the Snake River and its tributaries, concealing the incredibly violent volcanic history of the Yellowstone hotspot. Yet we now know that the hotspot created much of the flat expanse of the Snake River Plain. Like a boat speeding through water and creating an arc-shaped wave in its wake, the hotspot also left in its wake a parabola-shaped pattern of high mountains and earthquake activity flanking both sides of the Snake River Plain.

What Are Hotspots?

Earth is composed of layers of rock. The uppermost layer, known as the crust, is relatively thin beneath the oceans, sometimes only 6 miles thick. Under continents, however, the crust can range from 15 to 30 miles thick. The mantle is the layer beneath the crust, and is made of iron-rich rock that is more dense than the crust. The mantle extends to depths of about 1,800 miles. Beneath the mantle is Earth's molten outer core.

The lithosphere is the Earth's upper 50 to 80 miles, including the crust and uppermost mantle. The lithosphere is relatively brittle, and is broken into huge "plates" of rock, such as the North American plate and plates beneath the other continents and oceans. The widely accepted theory of plate tectonics says these plates slowly drift an inch or more each year. During Earth's 4.6-billion-year history, the lithospheric plates (often imprecisely referred to as crustal plates) drift and slide around the planet's surface, with continental plates sometimes joining to form giant supercontinents, sometimes breaking apart to form separate continents and intervening oceans.

Collisions of Earth's plates produce violent geological processes. The Pacific plate beneath much of the Pacific Ocean slides northwest past the North American plate about two inches a year. The boundary between these two plates is California's San Andreas fault zone, famed for producing destructive earthquakes. Farther north, the much smaller Juan de Fuca plate—located beneath the ocean offshore from Oregon, Washington, and British Columbia—is diving at an angle or "subducting" eastward beneath the North American plate. This occasionally produces gigantic earthquakes in the Pacific Northwest (the last one in 1700) and also is responsible for the formation of the Cascade Range and eruptions from its volcanoes, including Mount St. Helens, Mount Rainier, and Lassen Peak.

Tectonic plates are able to drift or slide because they float atop the asthenosphere, a part of the mantle in which warm rock acts like hot plastic and can slowly boil upward, flow horizontally, then sink downward again—a process called convection. This conveyor belt-like or treadmill-like flow of hot, malleable rock within the asthenosphere makes the overlying plates move. In a sense, Earth's plates are like the pieces of scum floating in pot of tomato soup, driven by the bubbling and boiling of the liquid underneath.

In the Earth, huge volumes of hot and molten rock rise upward in hotspots and beneath mid-ocean ridges, which extend thousands of miles along the seafloor. The rock rising from mid-ocean ridges and, to a lesser extent, from hotspots, drives the movement of tectonic plates. And much of what we know about the motions of Earth's plates was learned by studying the tracks left on them as they passed above hotspots.

Hotspots also are known as "mantle plumes." The traditional theory of hotspots is that they are columns or plumes of hot and molten rock that begin 1,800 miles underground at the boundary between Earth's core and lower mantle, then flow slowly upward through the entire mantle and crust. This theory is probably true for most hotspots. It also ultimately may prove correct for the Yellowstone hotspot. However, research in recent years suggests the Yellowstone hotspot may not have its roots at the core-mantle boundary, but instead originates at a shallower depth.

Medical CT scans bounce X rays through the human body to make three-dimensional pictures of internal tissues. In a similar manner, a method called seismic tomography uses hundreds of seismographs to measure the speed of seismic waves from quakes and small, intentional dynamite explosions—data that allow geophysicists to make three-dimensional pictures of structures within the Earth. The method works because seismic waves travel more quickly through cold, dense, solid rock than through hot, less dense molten or partly molten rock. Seismic tomography has been used to look at the geology of the crust and upper mantle beneath the Snake River Plain and Yellowstone National Park.

Such studies suggest the partly molten plume of the Yellowstone hotspot forms in the upper mantle, only about 125 miles beneath Earth's surface, and might not extend deeper. The seismic studies find no evidence of molten or partly molten rock below the Yellowstone hotspot at depths greater than 125 miles.

What makes rock melt in the upper mantle to form the roots of the Yellowstone hotspot? Some heat does rise from the core. However, radioactive decay and "decompressional melting" are likely sources of heat that melts rock at the roots of the Yellowstone hotspot. Concentrations of uranium and other radioactive elements in Earth's mantle produce large amounts of heat through radioactive decay, melting iron-rich basalt rock, which then rises buoyantly within the mantle. Other, less dense rock is left behind, and the pressure on that rock is reduced because the overlying basalt is moving upward. This reduces the melting point of the less dense rock, so it too begins to melt.

Hot and molten rock rises because it is less dense than cooler, surrounding rock. Beneath the sea, the iron-rich basalt often erupts from mid-ocean ridges, undersea boundaries between tectonic plates where new seafloor crust is produced. Eruptions from these "spreading centers" push the seafloor's tectonic plates away from the ridges in both directions, and the plates also are carried along by horizontal movement of the underlying mantle—truly a conveyor belt producing new seafloor. Indeed, North America drifts southwest over the Yellowstone hotspot because the entire continent is moving away from the mid-Atlantic ridge, located more than 5,000 miles east of Yellowstone.

Molten basalt also can rise in hotspots and erupt through overlying tectonic plates, producing undersea volcanoes that become islands if they are tall enough. As a seafloor plate moves, the underlying hotspot periodically erupts through weak zones within it, creating island chains like Hawaii noted for flowing eruptions of basaltic lava.

The process is somewhat different for the Yellowstone hotspot, which, unlike most other active hotspots, is located far from a plate boundary and beneath a continent instead of beneath the seafloor. The roughly 50-mile-wide plume of the Yellowstone hotspot slowly rises from its point of origin about 125 miles deep until it reaches a depth of about 50 miles, where it encounters the overlying North American tectonic plate (the uppermost mantle and crust), which is colder rock (Figure 2.1). There, the molten rock forms a pool perhaps 300 miles wide from side to side. The plume is even wider in the wake of the hotspot because it is dragged in the direction that the plate is moving, sort of like what would happen to honey if you poured it onto a table while someone was moving the table.

As the iron-rich basaltic magma pools at the base of the plate, big blobs of the molten rock periodically rise from the top of the hotspot plume and move upward through the rest of the upper mantle and into Earth's crust. The molten blobs of basalt heat overlying crustal rock, creating a "magma chamber" in which silica-rich crustal granite partially melts, forming a molten rock known as rhyolite when it erupts. (Silica, or silicon dioxide, is a glassy mineral that can take many forms, including sand, quartz, and opal. It is a common constituent of rocks in Earth's crust.) Because molten rhyolite is thick and viscous, major eruptions from the Yellowstone hotspot have been explosive, unlike the basalt that erupts more gently from oceanic hotspots.

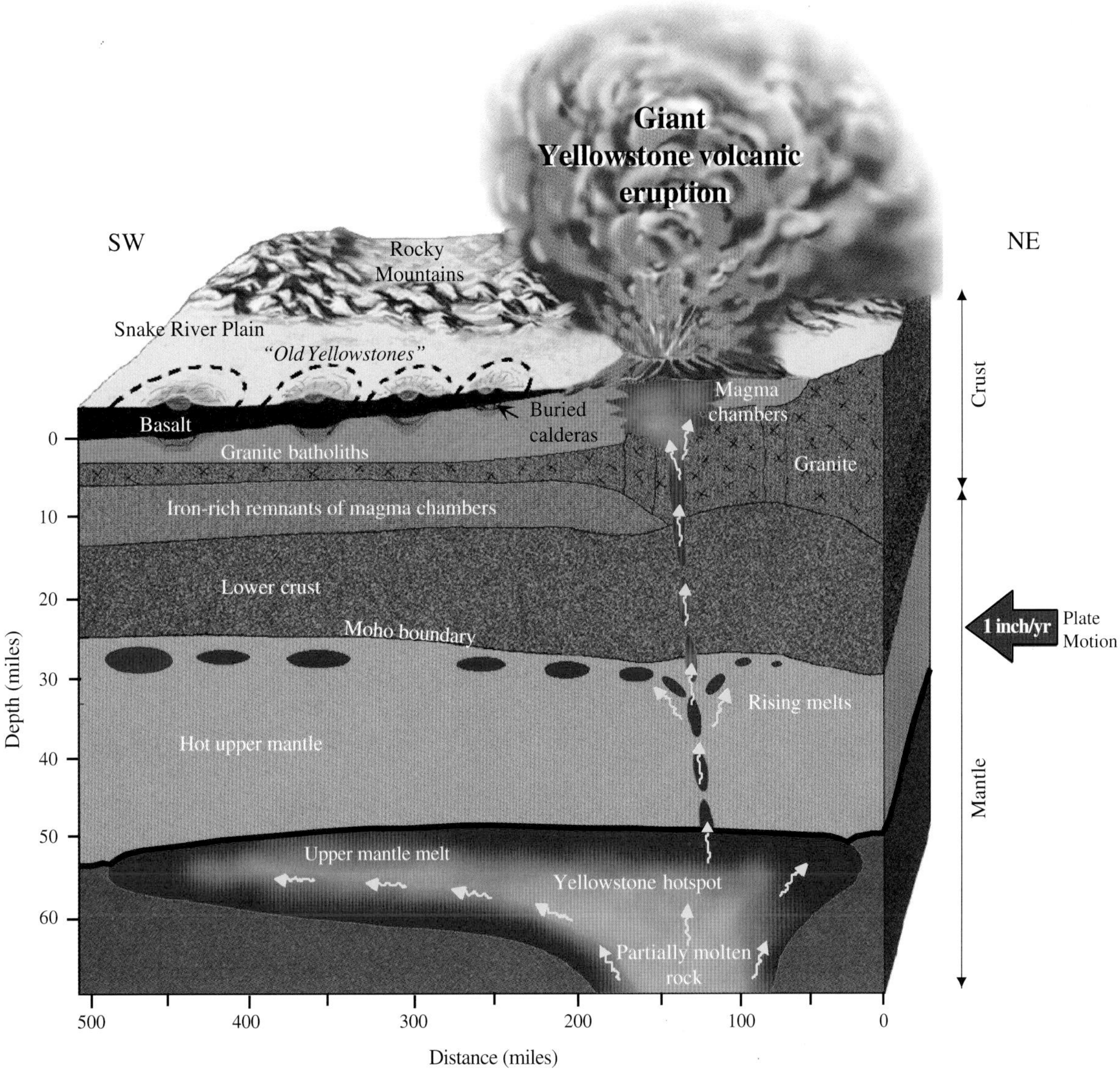

2.1 ⁓ *Cross section showing relationship of Yellowstone and the Snake River Plain to the Yellowstone hotspot and surrounding mountains. Molten or partly molten rocks called magmas (red) rise from the hotspot and are sheared off by the overlying North American plate, which moves southwest at 1 inch per year. Some molten rock rises upward, through the Moho, or crust-mantle boundary, melting overlying rock to create a magma chamber that fed Yellowstone's volcanic eruptions. Granite batholiths and iron-rich rocks beneath the Snake River Plain are remnants of old magma chambers left in the wake of the hotspot. Later basalt eruptions covered the Snake River Plain.*

The magma chamber beneath Yellowstone today is not a pool of completely molten rock. Instead, it is like a big, flattened, pancake-shaped sponge in which much rock is hot but solid and other rock is molten. The magma chamber is probably 5 to 8 miles beneath Yellowstone National Park. This magma chamber was responsible not only for prehistoric catastrophic volcanic eruptions at Yellowstone, but for smaller lava and ash eruptions as well. It also provides heat to power the park's geysers and hot springs.

A hotspot releases a large amount of heat from Earth's interior, altering the surface with extensive volcanic and geothermal activity.

Hotspots of the Solar System

Of the roughly thirty active hotspots on Earth, almost all except Yellowstone are beneath oceans or near coastlines or other boundaries between tectonic plates (Figure 2.2). The best known of the other hotspots are those that produced Iceland, the Hawaiian Islands, and the Galápagos Islands.

If you slowly move your hand over a candle flame, you will suffer a nasty burn along the path of the flame. As plates of Earth's surface slowly drift over hotspots, the overlying landscape is marked by eruptions of molten rock from below.

About 80 million years ago, the Hawaiian hotspot erupted through the Pacific plate, and periodically has done so ever since as the plate moved northwest at more than 3 inches a year. The eruptions of basalt lava built a 3,600-mile-long chain of atolls and undersea volcanoes, called the Empĕror-Hawaiian seamount chain, stretching from Kamchatka in easternmost Russia south and east through Midway Island to the Hawaiian Islands.

During the last 1.5 million years, eruptions from the Hawaiian hotspot produced volcanoes that rose above sea level to become the Hawaiian Islands. As the Pacific plate drifted northwest, the hotspot erupted repeatedly, producing a chain of islands extending from Kauai and Niihau on the northwest end of the chain to the Island of Hawaii, the youngest island at the southeast end. Hawaii and its famed volcanoes—Kilauea, Mauna Loa, and Mauna Kea—still sit atop the hotspot, which is why eruptions are frequent on the island, the site of Hawaii Volcanoes National Park. In addition to island-building eruptions, the Hawaiian hotspot makes a 900-mile-wide area of the Pacific seafloor bulge upward as much as 1,600 feet. The bulge or "swell" is centered beneath the Island of Hawaii.

Iceland is another notable hotspot. Unlike the Hawaiian hotspot, which punched through the overlying tectonic plate, Iceland is located on the mid-Atlantic Ridge, a

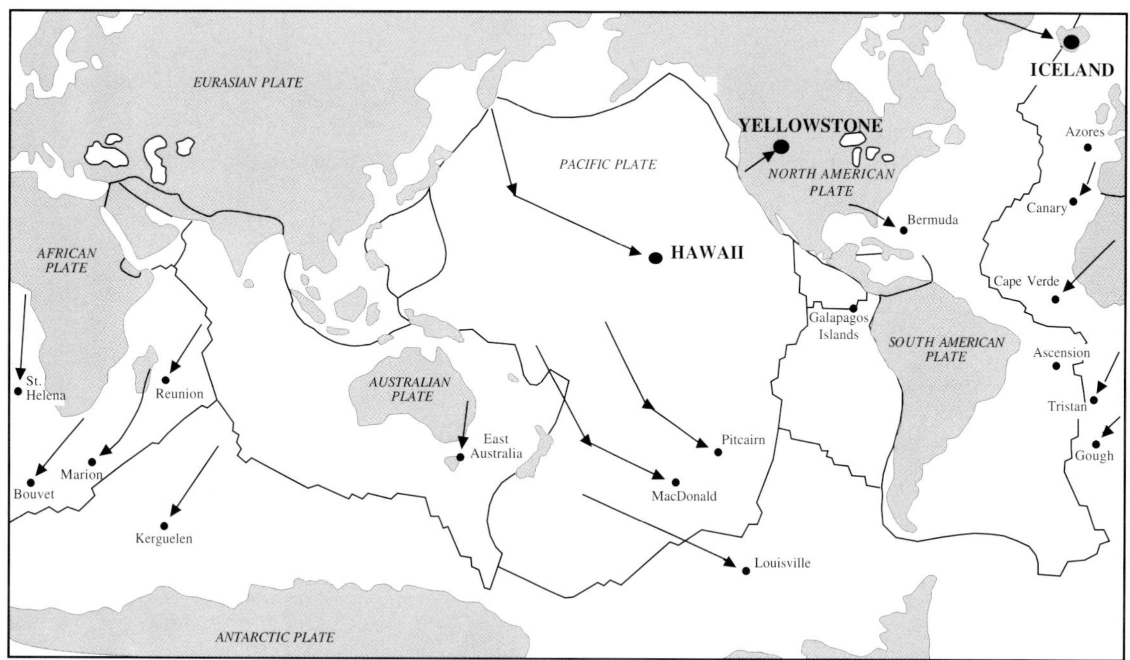

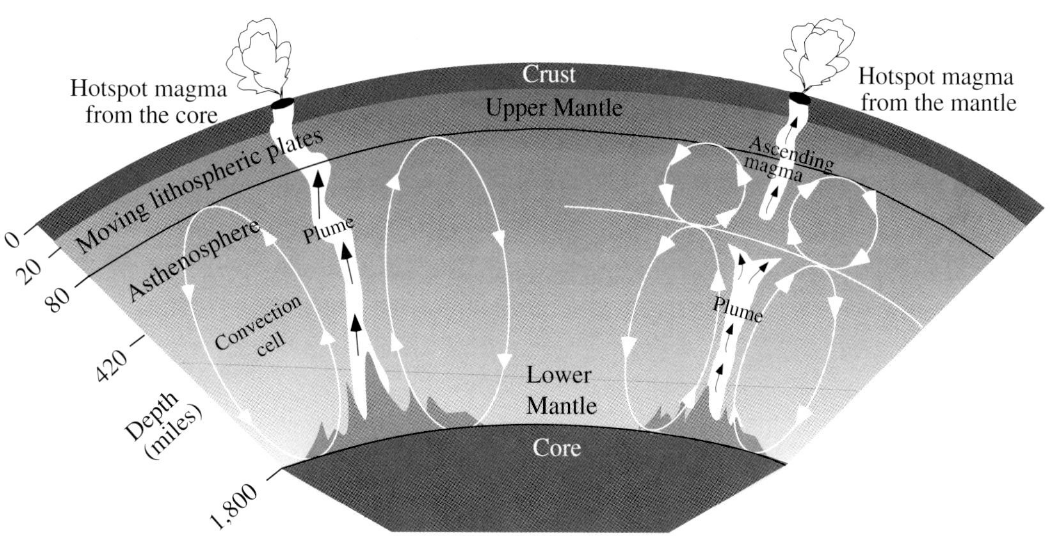

2.2 ~ Top: *Black dots are hotspots, areas of concentrated volcanism and heat derived from molten rocks that flow upward from Earth's interior. Arrows show motion of the hotspots as the plates of Earth's surface drift over them. Most hotspots, such as Hawaii and Iceland, are beneath oceans. The Yellowstone hotspot is under a continent.* Bottom: *Molten rock, or magma, rises in convection cells like water boiling in a pot. The left side shows the traditional view of a hotspot as a plume of magma rising from Earth's core. The right side shows the newer theory that the Yellowstone hotspot originates at a much shallower depth.*

roughly 12,000-mile-long spreading center from which magma erupts to create new seafloor, which is pushed east and west away from the ridge. While undersea eruptions occur along the length of the mid-Atlantic Ridge, the volume of eruptions from the hotspot is even greater, creating the island known as Iceland. The tiny nation's basaltic volcanism is similar to that of Hawaii, and its volcanoes erupt along well-defined fissures and vents above zones of weakness in the crust. These weak areas, literally cracks in the Earth, are produced because the hotspot beneath Iceland makes the island rise upward and spread apart sideways.

Research indicates the Iceland hotspot moved eastward across Greenland to Iceland as new ocean floor spread westward from the mid-ocean ridge during the past 70 million years. The hotspot plume is about 250 miles wide and starts in the mantle at least 250 miles beneath Earth's surface. One study found signs of a narrow Iceland hotspot plume at least 410 miles deep, implying the hotspot is rooted even deeper in the lower mantle, perhaps at the core-mantle boundary.

Another hotspot produced the Galápagos Islands, located in the eastern Pacific Ocean west of the South American nation of Ecuador. Some have suggested the same hotspot was responsible for large basalt eruptions that extended more than 1,400 miles from the Galápagos northeast through Costa Rica all the way to Curaçao Island in the Caribbean some 88 to 90 million years ago.

There are some extinct hotspots beneath Africa, and it has been suggested the higher-than-expected topography of southern and eastern Africa may represent a "superswell" that was lifted upward by hot rock rising from the core-mantle boundary. The rising rock may be in two young hotspots that have yet to break through to the surface and produce volcanic eruptions. Farther north, an active hotspot—the Afar plume—apparently is present beneath the south end of the Red Sea, between Ethiopia and the Arabian Peninsula.

Some researchers have argued an ancient hotspot once existed beneath southeastern Canada and the northeastern United States, and that another hotspot created a chain of volcanoes from Arizona east to New Mexico's Valles caldera, which erupted 1.2 million years ago, about the same time as the second of Yellowstone's three large caldera eruptions.

Yellowstone shares characteristics with oceanic hotspots: It makes the Earth bulge upward in a 300-mile-wide swell above it, and its roots involve iron-rich basalt rock melting in the mantle. However, Yellowstone is also fundamentally different. To trigger eruptions, the Yellowstone hotspot must push through much thicker, older, and colder continental crust, not the thinner, younger, warmer crust under the seafloor. Also, the continental crust is much more rich in silica than the basaltic oceanic crust.

Ocean hotspots punch to the surface with relative ease, generally during nonexplosive eruptions of basalt lava.

The Yellowstone hotspot must melt rocks in the much thicker continental crust. The molten basalt at depth melts overlying silica-rich rocks, particularly granites, which form rhyolite magma that is a million times more viscous—thick in consistency—than basalt magma. As rhyolite magmas rise, gas bubbles form within the molten rock. Because the rock is so viscous, tremendous gas pressures must accumulate before the bubbles can burst. That is why rhyolite eruptions like those from the Yellowstone hotspot are explosive, unlike the gentler Hawaiian eruptions.

Nevertheless, it is clear that hotspots beneath continents also can produce floodlike eruptions of basalt lavas. The final stage of Yellowstone hotspot volcanism beneath Idaho covered the Snake River Plain with basalt lava flows (see the last section of this chapter). Prehistoric hotspots also may have produced gargantuan floods of lava in Siberia, India's Deccan traps, and Ethiopia.

In recent years, scientists have realized that hotspots also occur elsewhere in our solar system, including on the planet Venus. That cloudy planet lacks the kind of moving tectonic plates that make up Earth's crust, but the National Aeronautics and Space Administration's *Magellan* spacecraft identified several hundred circular features known as coronae, measuring at least 120 miles in diameter, and smaller circular features called arachnoids. Both appear to be areas of the surface where a plume of rock once rose upward from the Venusian interior. In places, there are chains of these features. Because Venus lacks drifting plates, the implication is that the hotspot plumes themselves migrated over time. This and other recent evidence challenges the long-held assumption Earth's hotspots always are fixed features beneath moving tectonic plates. Venus' coronae and arachnoids may have been the source of unimaginable volcanic outpourings that literally repaved Venus' surface an estimated 300 million years ago.

Jupiter's moon Europa, which has a crust of ice over a deep ocean of liquid water, also is believed to have hotspots beneath its seafloor. Europa's hotspots, apparently caused by the decay of radioactive elements in the moon's core, may be responsible for warming the ocean so that its water periodically breaks through and gushes onto the frozen crust.

Io, another moon of Jupiter, is probably the most volcanically active body in the solar system, with volcanoes spewing sulfur dioxide hundreds of miles high and coating Io's surface with yellow sulfur. An infrared sensor on NASA's *Galileo* spacecraft identified about thirty hotspots on Io, many associated with eruptions and with temperatures of 1,340 degrees Fahrenheit. Indeed, the spacecraft's data suggests the temperature may reach 3,140 degrees Fahrenheit at one powerful eruption site—probably the hottest volcanism in the solar system. Researchers believe the intense volcanism shows that Io's crust is being recycled so rapidly that continents cannot form—something that may have happened early in Earth's 4.6-billion-year history.

On Mars, a plume of hot rock from the planet's interior may explain the origin of several large, extinct volcanoes in the Tharsis region, including Olympus Mons, which towers nearly 85,000 feet above the surrounding terrain, making it the tallest mountain in the solar system. NASA has said there is growing evidence hydrothermal activity was once widespread on Mars and may have played a role in fostering microbial life that possibly once existed on the Red Planet, just as hydrothermal activity is suspected of promoting early microbial life on Earth. NASA said it was "searching for Yellowstone on Mars."

Before the Yellowstone Hotspot

Most of this book deals with geological events within the past 20 million years: acceleration of Basin and Range stretching of Earth's crust in the West starting about 17 million years ago, eruptions from the Yellowstone hotspot in Nevada, Oregon, and Idaho during the past 16.5 million years, the birth and development of the modern Teton fault within the past 13 million years, and volcanic, seismic, and geothermal activity since the hotspot reached the site of Yellowstone National Park more than 2 million years ago. All these geological developments happened during a mere blink of an eye in Earth's history. The last 20 million years represent less than one-half percent of the 4.6 billion years of the planet's existence. What came before the relatively recent volcanic outbursts, earthquakes, glaciers, stream erosion, and landslides that sculpted modern Yellowstone, the Teton Range, and the Snake River Plain?

Over the eons, as giant plates of Earth's crust drifted slowly over the globe, supercontinents formed and later broke apart, eventually taking the shape of modern continents. Ocean levels rose and fell. Great masses of land heaved upward and sank downward. The primeval landscape was reshaped countless times by many of the same processes seen today.

Most of Earth's history—from the beginning to 570 million years ago—is known as the Precambrian era. Precambrian rocks are found in the Yellowstone–Teton region, including northern Yellowstone and the cores of the Teton, Beartooth, Wind River, and Gros Ventre ranges. The oldest rocks—3.4 to 3.8 billion years old—are found northeast of Yellowstone in the Beartooth Range and to the southeast in the Wind River Range. The heart of the Teton Range is also extremely old. It includes

2.8-billion-year-old granitic rocks in the east front of the range and 2.4-billion-year-old rock capping the central Teton Range. Precambrian rocks that are now thousands of feet high in the Tetons and nearby ranges were not born at such lofty heights. The rocks formed when they were about 6 miles underground and later were uplifted.

Rocks 2.8 billion years or older in the Beartooth, Wind River, and Teton ranges are believed to be the uplifted roots of primeval mountains that once extended from Wyoming northeast through Montana before they eroded. The 2.4-billion-year-old rocks in the Tetons are remnants of the roots of a somewhat more recent but still ancient mountain range that also was leveled by erosion. Vertical black walls or "dikes" of rock exposed in the central Tetons were formed when molten rock intruded into older Precambrian rocks about 1.5 billion years ago.

The mountains of the Precambrian slowly eroded and the elevated continental landmass subsided. So, during most of the Paleozoic era, from 570 million to 245 million years ago, the ebbs and flows of ancient seas repeatedly flooded the West. The Yellowstone–Teton region was often under ocean water, offshore from the western edge of primeval North America, which was part of a much larger supercontinent for much of the Paleozoic. The continent's west coast probably ran south to north through what are now Colorado, Wyoming, and Montana.

Yellowstone and Grand Teton national parks contain rock formations deposited as deep seafloor sediments during the Paleozoic, including limestones, sandstones, and shales. The ancient sediments are particularly well displayed on the north side of Yellowstone and the south end of the Teton Range. These formations range from the 535-million-year-old Cambrian period Flathead sandstone exposed atop Mount Moran in the Tetons to the 360- to 320-million-year-old Mississippian period limestones that produce dramatic cliffs seen northwest of Yellowstone in the Madison Range, in Hoback Canyon southeast of Jackson, Wyoming, and in the Snake River Canyon southwest of Jackson. Layers of Paleozoic rocks in the Tetons have been uplifted thousands of feet and tilted westward by movements along the Teton fault during the past 13 million years and to some extent by movements along an ancestral version of the fault as long as 34 million years ago.

The Mesozoic era extended from 245 to 66 million years ago and encompassed the Triassic, Jurassic, and Cretaceous periods. At times the Yellowstone–Teton region was covered by the ocean; at other times it was land, including sand dunes, vast tidal flats, and plains. During the Triassic period, iron-rich reddish siltstones and shales accumulated in swamps and shallow, muddy seas to the east of modern-day Yellowstone and the Tetons. Dinosaurs appeared during the Triassic and vanished at the end of the Cretaceous. Despite a limited amount of Mesozoic rock formations in the region, there are some dinosaur fossil sites around Yellowstone and Jackson Hole, so dinosaurs roamed the area during this era.

During the Cretaceous period, which started 144 million years ago, the central third of what is now the United States was flooded by a great, shallow sea. The water stretched from near the present site of the Mississippi River all the way west and north through Texas, Oklahoma, Nebraska, eastern New Mexico, Colorado, eastern Utah, the Dakotas, central eastern Montana, and most of Wyoming, including the Yellowstone–Teton region. The Cretaceous sea left thick deposits of shales, sandstone, and other rocks, including organic-rich sediments transformed by heat and pressure into coal and other hydrocarbons.

Later in the Cretaceous, western North America was squeezed in a roughly east–west direction, causing two great episodes of mountain-building that displaced the great seaway and ultimately built the Rocky Mountains. The West was compressed by the westerly motion of the North American plate pushing against the Pacific plate. The first mountain-building episode, known as the Sevier Orogeny, started roughly 110 million years ago. Large chunks of Earth's crust arched upward into folds or were thrust eastward along low-angle faults, starting in Nevada and Utah and eventually moving into Idaho and western Wyoming. About 80 million years ago, the squeezing continued and shifted eastward as the Laramide Orogeny began, uplifting the Rocky Mountains in Canada, Montana, Wyoming, Idaho, Utah, Colorado, and New Mexico. Part of the giant mountainous plateau of the young Rockies contained what later became western Wyoming's Teton, Gros Ventre, and Wind River ranges.

The uplift of the Rockies continued into the Cenozoic era, which represents the last 66 million years of Earth's history. It was a time of widespread mountain-building, volcanism, faulting, and glaciation that sculpted the Yellowstone–Teton region. Around 50 million years ago, during the Eocene epoch, numerous volcanic eruptions occurred on the north and east sides of Yellowstone, forming the Absaroka Range. Volcanic debris buried forests. Their fossilized remnants can be seen today at a fossil tree exhibit and along Specimen Ridge in northern Yellowstone. Like the modern Cascade Range in Washington and Oregon, the Absaroka volcanoes likely formed because the plate of Earth's lithosphere beneath the Pacific Ocean dove or "subducted" eastward beneath North America as the plates collided. Intermittent but subdued volcanism continued in the Yellowstone–Teton region until the hotspot reached the area some 2 million years ago and volcanism intensified.

Starting about 30 million years ago, much of the interior West began stretching apart in an east–west direction. The extension of Earth's crust in the region accelerated about 17 million years ago and continues today. This produced the modern Basin and Range topography—elongated north–south mountain ranges separated by long, north–south valleys—that extends from eastern California, Nevada, and southeast Oregon east into Utah, Idaho, and western Wyoming. The stretching affected the Teton region about 13 million years ago to begin the modern uplift of the Teton Range along the Teton fault. (There is evidence, however, that an ancestral version of the Teton fault was born as long as 34 million years ago when the West still was being squeezed by mountain-building forces.) Basin and Range extension of the West also created other faults in the Yellowstone–Teton region, including the Madison, Gallatin, Red Mountain, and South Arm faults.

About 16.5 million years ago, soon after Basin and Range widening of the West accelerated, a great period of volcanism began near what is now the meeting place of Nevada, Oregon, and Idaho. The Yellowstone hotspot reared its hot head.

Origin of the Yellowstone Hotspot

No one really knows when or where the Yellowstone hotspot was born. Some scientists have argued the hot and molten rock first rose toward Earth's surface because of a zone of weakness in Earth's crust and mantle at what is called the Mendocino Triple Junction. Now located along the northern California coast, the junction is a weak zone in Earth's crust where three tectonic plates intersect: the Pacific and Gorda plates beneath the ocean and the North American continental plate.

There also have been arguments that the Yellowstone hotspot originated even earlier, at least 60 million years ago, beneath the Pacific Ocean farther offshore from the triple junction. This theory says the hotspot originated beneath the Pacific plate of Earth's crust. As that plate dove or subducted eastward beneath North America, the continent essentially rode over the hotspot.

Some have argued the hotspot was born 23 to 50 million years ago when the junction was much farther south on the California coast and the entire Pacific coast much farther east (closer to the Rockies) than it is today. No volcanic rocks have been found to support the idea that the hotspot may have first erupted in primordial California. Some researchers, however, have argued volcanic rocks found in the Yukon erupted from the Yellowstone hotspot about 70 million years ago and that, at the time, the hotspot was beneath the present location of the Oregon coast. The rocks later were carried north to the Yukon atop the Kula plate of Earth's lithosphere—a plate once located beneath the Pacific Ocean before it dove beneath Alaska and the Yukon and was destroyed.

Another theory is the hotspot may have first appeared 40 million years ago near the California–Nevada border, and that its rising heat and molten rock played a key role in making the overlying crust spread apart starting 30 million years ago—the first stage of Basin and Range stretching of the West. The east–west pulling apart of the region has doubled the distance between Reno, Nevada, and Salt Lake City, Utah, during the past 30 million years. Some evidence supports the notion that the Yellowstone hotspot initiated Basin and Range stretching of Earth's crust. The evidence is a large body of low-density rock under western Nevada. It might be the cooled and solidified remnant of the Yellowstone hotspot's original plume head.

The earliest definitive evidence of the hotspot's existence dates to about 16.5 million years ago. Around that time, the hotspot produced its first cataclysmic caldera eruptions in northern Nevada and southeast Oregon. No one really knows why the hotspot burped its way to the surface there. Some have speculated the hotspot erupted after a large meteorite punched a hole in Earth's crust near the Oregon–Nevada border. A much more likely scenario, however, ties the hotspot to the stretching apart of the Basin and Range Province. The rising plume of the hotspot not only may have helped initiate such crustal extension 30 million years ago, but also may have contributed to the speedup of such stretching 17 million years ago in central Nevada. That is not far from the hotspot's first known caldera eruptions near the Nevada–Oregon–Idaho border 16.5 million years ago.

Traditional hotspot theory—the notion of a plume of molten rock rising all the way from Earth's core—implies that the landscape above has no effect on where hotspots form. Instead, hotspots would form where hotter or molten rock rises due to variations in rock temperatures or densities. The traditional theory is backed by research showing large areas of partially molten rock at the core-mantle boundary deep beneath active or extinct hotspots in Iceland, the Azores, East Africa, and the south-central Pacific. Volcanic rocks on Earth's surface above many hotspots also have a geochemical makeup indicating they originated deep in the mantle and perhaps at the core-mantle boundary.

The newer theory—that at least the Yellowstone hotspot has roots only 125 miles deep—suggests its formation can be related to the overlying landscape, with heat and molten rock rising in weak zones such as the Basin and Range Province where Earth's crust is extending and getting thinner, like taffy being pulled apart.

If the hotspot's first known eruptions were related to the stretching of Earth's crust, as seems likely, it is possible the two processes enhance each other. On one hand, the rising plume of hot and molten rock would have made the Oregon–Nevada–Idaho border area bulge upward, possibly accelerating the rate at which Earth's crust stretched apart in the region. On the other hand, Earth's crust and upper mantle already were being stretched in the area. Such extension made Earth's crust thinner and could have eased the pressure above the hotspot, accelerating the ascent of basaltic magma from the roots of the hotspot. The molten basalt then heated and melted overlying granites, producing the rhyolite magma that ultimately erupted catastrophically to the surface, forming Yellowstone-style calderas 16.5 million years ago in northern Nevada.

Some researchers have argued that a few million years before the hotspot began erupting near the Oregon–Nevada border, the head or top of the hotspot plume began spreading laterally at the base of the lithosphere, and a large blob of molten basalt slowly leaked upward toward the northwest, erupting between 17 and about 14 million years ago to cause the massive initial phase of the Columbia River basalt lava flows that literally flooded the Pacific Northwest until about 6 million years ago. The timing of the early Columbia River flood basalts comes amazingly close to the speedup of Basin and Range extension of the West 17 million years ago and Yellowstone hotspot's first eruptions near the Nevada–Oregon border 16.5 million years ago. It is tempting to theorize the three events are all tied to the hotspot. Yet some scientists see no link between the Yellowstone hotspot and the Columbia River basalt flows. They argue those eruptions were caused by the same forces that created the Cascade Range volcanoes: the plate of Earth's crust beneath the Pacific Ocean diving or subducting beneath the Pacific Northwest, then melting to produce magma for volcanic eruptions.

Although the Yellowstone hotspot's connection to Columbia River volcanism remains speculative, there is no doubt the hotspot completely reshaped the landscape of southern Idaho and northwest Wyoming. As the North American plate drifted southwest over the hotspot, the hotspot kept blowing up, producing about 100 gigantic caldera eruptions during the past 16.5 million years. Because the plate moved southwest, progressively younger calderas line up in a chain stretching some 500 miles from the Nevada–Oregon border northeast up Idaho's Snake River Plain to the modern site of Yellowstone in Wyoming.

At the same time, Basin and Range stretching in Idaho and western Wyoming has been primarily in a northeast–southwest direction, in contrast to Nevada and western Utah, where the crust pulls apart in an east–west direction. The northeast–southwest widening of Earth's crust in Idaho paralleled the southwest movement of the North American plate and the resulting northeast movement of the Yellowstone hotspot. The pulling apart and thinning of Earth's crust may have decreased pressures to allow molten rock to erupt to the surface, not only for the hotspot's first caldera-forming eruptions along the Nevada–Oregon border, but for the whole series of caldera eruptions stretching across Idaho and into Yellowstone.

Outpourings from Hell

Caldera-forming eruptions are unrivaled among volcanic eruptions, and no eruption of such size has happened during recorded history. They expel hundreds to thousands of times more magma—underground molten rock—in a shorter period than any other kind of eruption. Some prehistoric basalt flows on continents did have larger volumes. So do mid-ocean ridges—the sites of underwater volcanic vents that continuously erupt to produce new seafloor. But basalt flows and undersea volcanoes erupt over longer periods, not in a sudden catastrophe.

Colossal caldera-forming eruptions are fed by magma chambers—the large, sponge-like bodies of molten and partially molten rock in the crust. These magma chambers are known as plutons, after Pluto, the Greek and Roman god of the underworld. Magma chambers can begin a few miles beneath the surface and extend to depths of 8 miles. They typically are a few tens of miles in diameter. The magma chamber now under Yellowstone National Park is believed to measure about 45 miles by 30 miles—the same size as the overlying caldera. When a pluton cools and solidifies, it is called a batholith.

As fluid and gas pressures build up in a rhyolite magma chamber, molten rock begins to move toward the surface around the perimeter of the chamber. Growing pressures push upward until finally the magma breaks through the overlying rocks, triggering earthquakes in the process.

When molten rhyolite erupts on the surface, the high pressures are released in a catastrophic caldera-forming eruption of awesome power. The initial rupture expands rapidly, blowing out huge volumes of superheated water vapor, gases, preexisting solid rock, and magma. This debris is hurled from the caldera at supersonic speeds. The magma, pulverized by the sudden expansion of trapped gas bubbles, explodes almost instantaneously into ash, pumice, and bigger rocks. The eruption propels a suffocating cloud of water vapor, gases, and ash 60,000 to 80,000 feet or more into the atmosphere.

Large lava flows erupt from the caldera. For perhaps 60 miles around the caldera, a fountain-like eruption plume rains hot ash and rock on the ground, where their heat welds them into layers of tuff as much as several hundred yards thick. At increasing distances, smaller and smaller particles rain to the ground. Even a few hundred miles away, ash may accumulate to thicknesses of several inches to a foot. Jet streams carry ash and volcanic gases around the globe for years, cooling the climate and creating flaming sunsets.

After the eruption, the remaining hole in the ground—the caldera—measures a few tens of miles wide and hundreds of yards deep. The caldera floor drops downward during the eruption along faults that form a giant ring inside the larger ring-shaped caldera rim. Mountains that once crossed the area sink into the caldera—if they have not already been blown away. The eruption began along the ring-shaped fault system. These faults roughly parallel the outer edge of the original, buried magma chamber. As the caldera floor sinks, it becomes uneven. Then, over the ensuing hundreds of thousands of years, smaller lava flows—first of explosive rhyolite, then later, nonexplosive basalt—erupt and spread across the caldera floor, smoothing its rough topography.

Unlike the Yellowstone caldera, the older calderas beneath the Snake River Plain have been covered by basalt flows, so much less is known about their explosions than about the last 2 million years of eruptions since the hotspot reached its present position beneath Yellowstone. We will examine the severe effects of those eruptions in the next chapter. However, like the later Yellowstone eruptions, the caldera-forming eruptions beneath the Snake River Plain must have been catastrophic. For example, one such eruption almost 12 million years ago dumped a foot of volcanic ash as far away as Nebraska, killing and burying herds of camels, rhinos, and horses that became fossilized and later were found by paleontologists.

Track of the Hotspot

Beneath the relatively recent basalt lava flows that cover the Snake River Plain and its ancient calderas are rocks made of lighter, silica-rich rhyolite, which was spewed out as ash and sticky lava during caldera-forming eruptions. These rocks—often tinged with orange or purple—can be seen in some places where the Snake River cut downward to create canyons. Such rocks also have been removed from holes drilled deep into the Snake River Plain.

The rhyolite rocks date to about 16.5 million years at the old, buried calderas near the Nevada–Oregon border. The same type of rocks are younger and younger as one moves northeast up the Snake River Plain and into Yellowstone. The progressive decrease in the age of rhyolite rocks provides the best evidence that a string of giant caldera eruptions progressed from the Nevada–Oregon border across Idaho to Yellowstone as the North American plate moved southwest.

Rhyolite rocks along the path of the hotspot have been dated and show 100 distinct ages. That indicates the hotspot triggered about 100 caldera eruptions as North America drifted over it. Some of the calderas erupted more than once, and other caldera-forming eruptions overlapped with older calderas. Depending on how you count such overlaps, the 100 caldera eruptions originated from at least seven "volcanic centers" (see Figure 1.3) stretching from the Nevada–Oregon border to Yellowstone:

- The McDermitt volcanic field, in northern Nevada near the Oregon border, erupted in multiple caldera-forming eruptions from 16.5 to 15 million years ago.
- The Owyhee volcanic center, which spans the Nevada–Oregon–Idaho border, erupted repeatedly between 15 and 13 million years ago.
- Calderas in the Bruneau–Jarbidge volcanic center, located southeast of Boise, Idaho, erupted several times between 12 and 10.5 million years ago.
- The Twin Falls volcanic center northeast of Twin Falls, Idaho, blew up repeatedly between 10.5 and 8.6 million years ago.
- The Picabo volcanic center's calderas, located near Blackfoot, Idaho, erupted from about 10 to perhaps 7 million years ago.
- The Heise volcanic center, located mostly north of Idaho Falls, produced at least three big caldera-forming eruptions 6.5, 6, and 4.3 million years ago.
- The Yellowstone volcanic center blew up three times, creating the original Yellowstone caldera 2 million years ago, the Island Park, Idaho, caldera 1.3 million years ago, and the modern Yellowstone caldera 630,000 years ago.

The dates of eruptions from the various volcanic centers can be used to calculate the speed at which North America drifted southwest over the hotspot. The average rate for the last 16.5 million years is about 1.8 inches annually. However, if shorter time intervals are analyzed, it turns out the plate moved about 2.4 inches per year from 16.5 million years ago until about 8 million years ago, then slowed to 1.3 inches a year. At about the same time the plate movement started slowing down, its movement toward the southwest turned a bit more southerly, which meant the path of the hotspot toward the northwest turned slightly more toward the north. The reason the speed and direction of plate motion changed is unknown. However, it explains why the Snake River Plain turns more to the northwest at a point east of Twin Falls.

The movement of the Yellowstone hotspot from the Nevada–Oregon–Idaho border was not entirely caused by the southwestward movement of the North American plate. At the same time the plate drifted southwest, Basin and Range stretching of Earth's crust pulled Idaho apart in a southwest–northeast direction at a rate of about a half inch per year. Thus, the hotspot's movement across Idaho and into Yellowstone was partly due to the drifting of the whole continent and partly to the stretching apart of the West.

Big Bulges

The hotspot not only triggered caldera-forming eruptions as it crossed under Idaho, it created a "topographic swell"—a broad area in which Earth's surface bulges upward. Such swells occur because rising hotspot magma pools at the base of the lithosphere, about 50 miles underground. The molten rock is buoyant, and literally uplifts the overlying upper mantle and crust to make Earth's surface bulge upward. Swells are common above oceanic hotspots, which can make the seafloor bulge upward as much as 3,000 feet. Seafloor swells can be 400 to 900 miles wide.

Yellowstone National Park and a large area surrounding it sit atop a swell roughly 1,700 feet tall and about 300 miles wide. The current site of Yellowstone was already high in the Rocky Mountains tens of millions of years before the hotspot arrived. But if it was not for the bulge created by the hotspot, the modern Yellowstone Plateau would be at an elevation above 6,000 feet instead of its actual elevation of about 8,000 feet. The extra elevation caused by the swell contributed to the formation of a huge icecap that covered Yellowstone with as much as 3,500 feet of ice during the Ice Age.

As North America moved southwest over the hotspot, it first lifted the land above it, then caused caldera-forming eruptions, followed by smaller rhyolite lava eruptions and finally a blanket of basalt lava flows. So the Snake River Plain was once higher than it is today. As the hotspot and the bulge above it moved northeast, the Snake River Plain sank roughly 2,000 feet in its wake.

The sinking of the Snake River Plain in the wake of the hotspot created the eastern two-thirds of the vast valley (not the western third near Boise). The eastern part of the plain then captured the Snake River and its tributaries. Basalt lava flows—the last stage of hotspot volcanism—contributed nutrients to the rich soils of southern Idaho.

Quakes in the Wake

Faults flanking the path of the hotspot have made Yellowstone and the mountains on both sides of the Snake River Plain among the most seismically active areas in the West. Since 1900, the region has been rocked by thirty moderate to large quakes—those with magnitudes above 5.5.

If you imagine the Yellowstone hotspot as a boat that moved northeast across Idaho to Yellowstone, the left side of its "wake" is a belt of mountain ranges and earthquake epicenters on the north side of the Snake River Plain, stretching from Yellowstone west–southwest across south–central Montana and central Idaho. The magnitude-7.5 Hebgen Lake quake fell along this belt of seismicity, as did the magnitude-7.3 Borah Peak earthquake of 1983, farther west in Idaho. That jolt—which killed two people in Challis, Idaho—was the region's second strongest recorded quake.

The right side of the hotspot's "wake" is a band of mountains and quake faults extending from Yellowstone south–southwest through Wyoming and southeast Idaho to the northernmost part of Utah's Wasatch Range and Wasatch fault. This band of mountains and seismicity includes the Teton fault and constitutes the central part of the Intermountain Seismic Belt. The belt is a zone of earthquake activity that runs from southern Nevada northeast and then north through Utah, then northeast along the Idaho–Wyoming border to the Teton fault and Yellowstone, and then west–southwest, cutting across south–central Montana and into northern Idaho.

The hotspot did not create the mountains in its wake. Its caldera eruptions simply destroyed mountains that once crossed the Snake River Plain, while leaving mountains on both flanks of the plain intact. The hotspot also did not create the quake faults in the surrounding mountains, but destroyed faults that once crossed the Snake River Plain. Indeed, the Intermountain Seismic Belt at one time probably took a more directly northerly path from Utah through Idaho and into Montana. But as the hotspot repaved the Snake River Plain and blew away shallow faults, the seismic belt was deflected eastward. That is why, after running north through Utah, the Intermountain Seismic Belt bends northeast to the Teton fault and appears to wrap around the track of the Yellowstone hotspot.

The Snake River Plain itself has relatively few quakes. Some scientists have argued the crust is warmer beneath the plain than under surrounding mountains, and that the warmer rock is more flexible and less prone to rupture and produce quakes. Another theory is that the remnants of old magma chambers beneath the Snake River Plain contain extremely dense rock that is rigid and resistant to rupturing during quakes.

The Hotspot's Last Stage: Burps of Basalt

Seismic tomography has been used not only to look at the roots of the hotspot now beneath Yellowstone National Park, but to examine the rocks left beneath the Snake

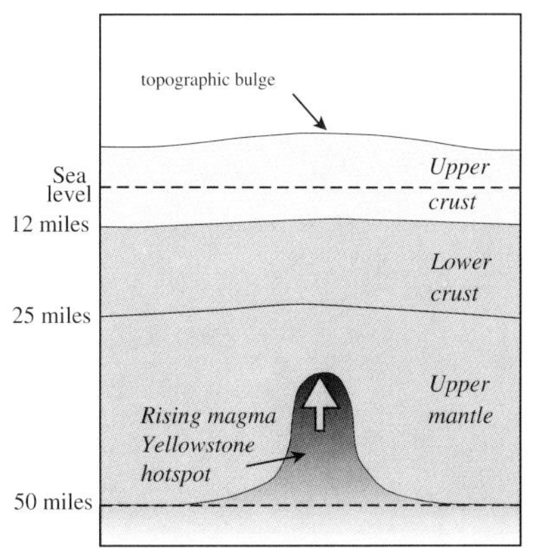

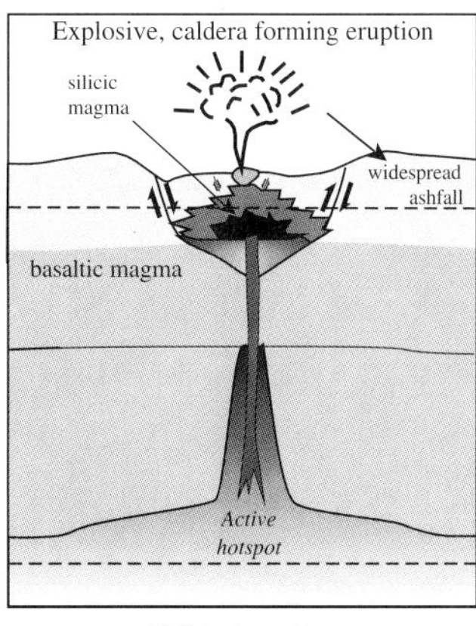

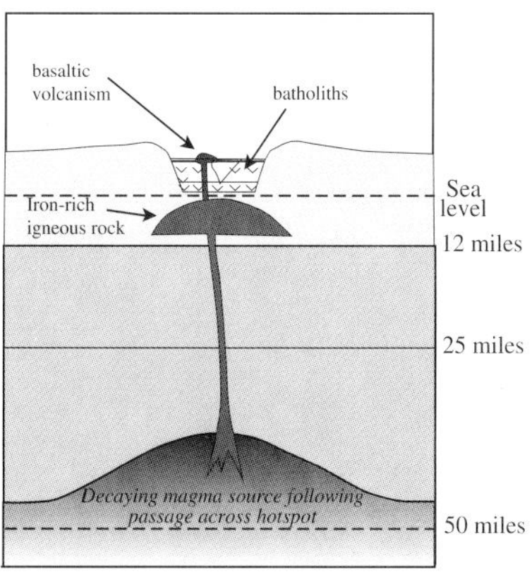

2.3 ~ *Stages of Yellowstone hotspot volcanism. During the "pre-Yellowstone stage," molten basalt rock rises from depth to 50 miles, making a broad area of the overlying landscape bulge upward as much as 1,700 feet. During the "Yellowstone stage," shallower molten rhyolite lava erupts during catastrophic caldera-forming eruptions and smaller eruptions. During the "Snake River Plain stage," the ground cools and sinks downward after the landscape had moved past the hotspot. Rhyolite eruptions cease, but darker basalt lavas erupt and cover the earlier rhyolite lavas and ashes. Modern Yellowstone National Park has gone through the first two stages; the Snake River Plain, all three.*

River Plain in the wake of the hotspot. The technique shows that beneath pockets of superficial sediments, the Snake River Plain is covered by a surface layer of iron-rich basalt lava roughly a mile thick—thicker in some places, thinner in others. Beneath that is a layer of lighter-colored, less dense, silica-rich rhyolite that is about a mile or two thick. That layer, in turn, is underlain by a 6- to 9-mile-thick layer of extremely dense rock that is even more iron-rich than the basalt on the surface. These layers of rocks are a record of the three stages of volcanic processes triggered by the hotspot beneath the Snake River Plain as the hotspot moved slowly northeast toward Yellowstone (Figure 2.3).

During the first or "pre-Yellowstone stage," heat from radioactive decay and reduced pressure in the crust allow basaltic rocks to melt and form the roots of the hotspot about 125 miles beneath Earth's surface. The red-hot basaltic magmas rise through the partly molten hotspot plume and hit the base of the lithosphere, about 50 miles underground, where the molten basalt ponds to form the top or head of the plume. This process makes the ground above bulge upward over a roughly 300-mile-wide area.

The second or "Yellowstone stage" begins when huge blobs of dense, molten basalt are sheared off the top of the hotspot plume by the movement of the overlying tectonic plate (the lithosphere, or crust and uppermost mantle). The molten basalt blobs start rising into the crust, primarily because they are less dense and more buoyant than surrounding solid rock and partly because the crust is being stretched apart by Basin and Range extension of the West. As the molten basalt blobs rise into the crust, they melt some of the surrounding silica-rich granite, which is known as rhyolite when it later erupts. The melting creates a large magma chamber known as a pluton, which measures a few tens of miles in diameter and is roughly 3 to 8 miles underground. The chamber is partly molten and partly solid, with molten rock flowing somewhat like water in a sponge.

Upward pressure from the magma chamber makes the crust immediately above it bulge upward—a smaller, localized bulge atop the broader regional swell or bulge. The bulging of the overlying crust accelerates earthquake activity, both from new fractures and pre-existing faults.

The "Yellowstone stage" reaches a climax when faults encircling the bulge produce earthquakes and the faults break deeper and deeper into the crust until they intersect the magma chamber. This allows molten rhyolite to flow to the surface along a series of faults that form a ring or circle a few tens of miles wide when viewed from above. As the sticky or viscous rhyolite magma rises and the pressure of overlying rock diminishes, gas trapped within the magma expands rapidly. The magma erupts explosively, spewing hot rocks and creating a huge eruption plume of volcanic ash and gas that is hurled high into the atmosphere at supersonic speeds. Some of the rhyolite magma spews from the ground to form fast-moving, superhot "pyroclastic flows" that incinerate anything in their path. As the eruption proceeds and the underground magma chamber is drained, what is left of the ground above it sinks downward, creating a caldera, or huge crater the same diameter as the underground magma chamber. Volcanic ash and gas in the atmosphere are carried by winds around much of the Earth. Ash piles up inches to a foot deep even hundreds of miles from the caldera. Closer to the caldera, the ground is blanketed with hot ash as much as several yards deep. The weight and heat of the ash make this layer fuse into a solid rock known as welded tuff.

Such catastrophic caldera eruptions can occur repeatedly in one area over the span of a few million years as long as the hotspot remains underneath the area. Each huge caldera eruption is followed by smaller flows of rhyolite lava—known as post-caldera eruptions—that eventually fill in the caldera over tens to hundreds of thousands of years.

The third or "Snake River Plain stage" of hotspot volcanism occurs as the North American plate continues to drift southwest, eventually cutting the connection between the caldera and the hotspot. The rhyolite magma chamber cools, allowing deeper basalt magma to ascend to the surface and erupt. Because the molten basalt contains less gas and is less viscous than molten rhyolite, it erupts nonexplosively, producing gentle flows of basalt lava that flood the landscape over millions of years. The basalt lavas eventually smooth the topography by covering the tuff and rhyolite rocks deposited during caldera and post-caldera eruptions.

Meanwhile, as the hotspot migrates away, the crust cools and contracts, and the broad regional bulge in Earth's surface sinks downward in the hotspot's wake. The surface also sinks because the rock left in the magma chamber is extremely dense, and its weight plus dense rocks in the upper mantle literally pull the overlying crust downward. As the hotspot moves over millions of years, the depressions in its wake—basically "Old Yellowstones"—form a broad valley.

How do the hotspot's volcanic stages relate to the three rock layers beneath the Snake River Plain? The lowest of the three layers—the extremely dense iron-rich rock—is made of the slag-like remnants of extinct magma chambers. The lighter molten rhyolite that once occupied the magma chamber erupted away, followed by basalt eruptions, leaving a slag of dense rock that did not erupt. The next layer up is the lighter rhyolite tuff and lava from caldera and post-caldera eruptions. And the surface layer of basalt comes from the gentle basalt lava flows that covered each old caldera as the hotspot moved away from it and progressed toward Yellowstone. The total volume of these basalt flows is probably several times greater than the volume of all the eruptions at Yellowstone since the hotspot arrived there 2 million years ago, but the Snake River Plain basalts flooded the area over a much longer period beginning sometime after 16.5 million years ago.

The three layers—dense slag, lighter rhyolite, and surface basalt—extend from the Nevada–Oregon border up the Snake River Plain as it crosses Idaho, although the surface layer of basalt lava becomes progressively thinner closer to Yellowstone. At Yellowstone National Park itself there are not three layers.

The bottom layer of heavy slag in dead magma chambers under the Snake River Plain corresponds to the partly molten and still active magma chamber beneath Yellowstone. On the Snake River Plain, the rhyolite layer is sandwiched above the slag and below the more recent basalt flows, but at Yellowstone, the rhyolite layer is on the surface, the product of catastrophic caldera eruptions and post-caldera rhyolite flows during the past 2 million years. And finally, while there have been a few small basalt flows at Yellowstone, the park largely lacks the top layer of basalt lava flows that cover the Snake River Plain. That is because extensive basalt flows represent the third and final stage of hotspot volcanism. Yellowstone National Park has not reached that stage because it still sits atop the hotspot and could blow up again in a giant caldera eruption—just as it did three times since the hotspot reached Yellowstone about 2 million years ago.

Cataclysm! The Hotspot Reaches Yellowstone

3

Epicenters from numerous earthquakes fall approximately along two parallel lines that stretch from southeast to northwest through Yellowstone National Park. During the past 630,000 years, lava flowed from eruptive vents located roughly along the same lines. The alignment of earthquakes and small volcanoes suggests that zones of weakness are deep beneath them within the Earth. Those zones may be the still-active roots of faults that once ran along the base of towering mountains.

Such mountains would have made ancient Yellowstone resemble today's Grand Teton National Park. Indeed, a few million years ago these mountains may have stretched northward through Yellowstone and hooked up with the Gallatin Range, which now extends from Montana south into Yellowstone's northwest corner.

So why is today's Yellowstone Plateau relatively flat? What happened to the mountains that once may have rose thousands of feet skyward like the Tetons do today? The answer, quite simply, is that they were destroyed 2 million years ago during a caldera eruption, which is the largest, most catastrophic kind of volcanic outburst—an explosion so cataclysmic that it dwarfs any eruption in historic time.

North America had continued its southwestward slide over the Yellowstone hotspot. After blasting and repaving the Snake River Plain, the hotspot was finally be-

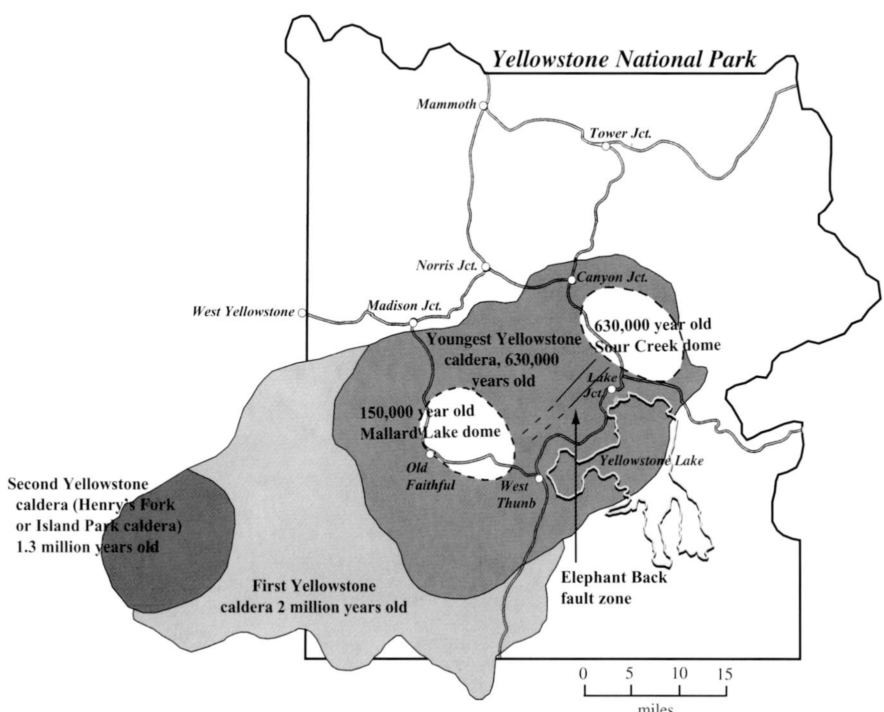

3.1 ⁓ *Yellowstone's main volcanic features include 2-million-, 1.3-million-, and 630,000-year-old calderas as well as the Mallard Lake and Sour Creek resurgent domes, which may be connected by the Elephant Back fault zone.*

neath the place for which it later was named. The power of its rising heat and hot rock began to shape Yellowstone into what it is today.

The first eruptive blast at Yellowstone 2 million years ago left a gigantic hole in the ground—a hole larger than the state of Rhode Island. The huge crater, known as a caldera, measured about 50 miles long, 40 miles wide, and hundreds of yards deep. It extended from Island Park in Idaho to the central part of Yellowstone in Wyoming (Figure 3.1).

During the volcanic cataclysm, hot ash and rock blew into the heavens over Yellowstone, then rained like hell from the sky. As heavier pumice and ash particles debris piled up on the ground, their heat welded the debris together to form a layer of solid rock called ash-flow tuff or welded tuff. This layer of rock—named the Huckleberry Ridge tuff—was hundreds of yards thick and was deposited for roughly 60 miles or more around the caldera.

Lighter ash particles covered the western half of North America, likely a foot deep several hundred miles from Yellowstone and inches thick farther away (Figure 3.2). Wind carried the lightest ash particles around the planet. For years after the eruption 2 million years ago, Earth's climate must have been chilled a bit as vast amounts of volcanic ash and gas filled the skies.

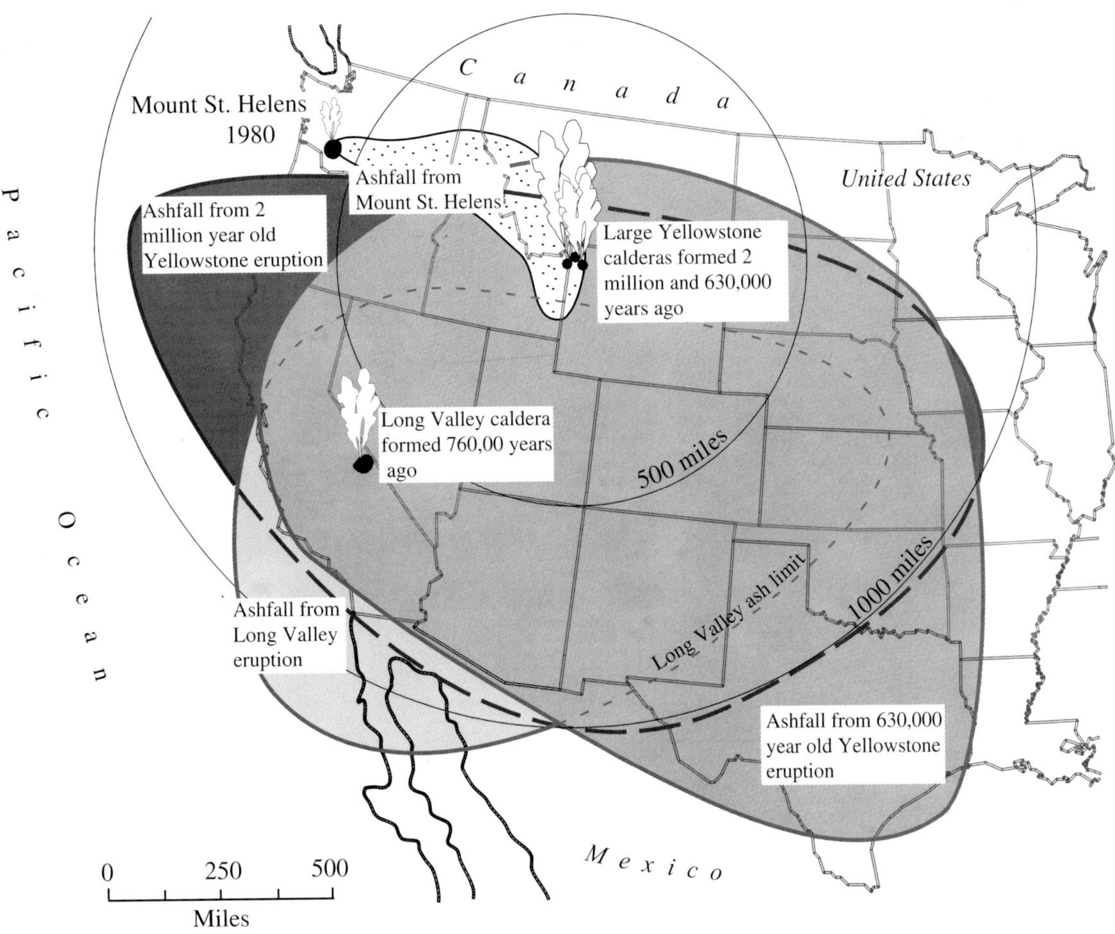

3.2 ∼ *Areas of the United States that once were covered by volcanic ash from Yellowstone's giant eruptions 2 million and 630,000 years ago, compared with ashfall from the 760,000-year-old Long Valley caldera eruption at Mammoth Lakes, California, and the 1980 eruption of Mount St. Helens, Washington. (Adapted from Sarna-Wojcicki, 1991.)*

The scale of the eruption was almost unimaginable. To get an idea of its immensity, consider Washington state's Mount St. Helens, which blew up on May 18, 1980, killing fifty-nine people, flattening surrounding forests, sending mudflows down nearby rivers, and dumping several inches of fluffy, gritty ash over wide areas of eastern Washington. The total volume of ash and other debris that erupted from Mount St. Helens was almost one-quarter of a cubic mile—equal to a giant cube of rock six-tenths of a mile on each side.

The Yellowstone caldera explosion 2 million years ago ejected an estimated 600 cubic miles of lava and ash—enough to fill a cube measuring 8.4 miles on each side.

So the eruption was 2,500 times larger than the Mount St. Helens disaster. That is a minimum. Estimates of the volume of material from the Yellowstone eruption are based on the amount of solid Huckleberry Ridge tuff deposited by the eruption and on estimates of ash fall. Mount St. Helens spewed mostly ash, which is less dense than tuff. If this is taken into account, the Yellowstone eruption 2 million years ago may have been more than 8,000 times larger than the 1980 Mount St. Helens explosion.

If all the ash, lava, and rock from Yellowstone's caldera eruption were piled up on Washington, D.C., the city would be buried by a layer of rock 10 miles deep. The same amount of material erupted from the caldera would bury the entire state of New York 67 feet deep. Wyoming would be buried 38 feet deep. Even spread over California, the volcanic material would be 20 feet deep.

As if that wasn't enough, the cataclysmic blast at Yellowstone 2 million years ago wasn't the last, although it was the biggest. Heat and molten rock from the hotspot shifted somewhat west, and 1.3 million years ago another gargantuan eruption blew a 15-mile-wide hole in the ground known as the Henry's Fork or Island Park caldera. That eruption blew out 67 cubic miles of hot ash and rock—a blast 280 times larger than Mount St. Helens's eruption.

Then, 630,000 years ago, catastrophe struck again, blasting out a new 45-by-30-mile-wide caldera that overlapped the 2-million-year-old caldera and was located more to the east. That catastrophe was 1,000 times bigger than the 1980 Mount St. Helens eruption, spewing out 240 cubic miles of hot rock, ash, and other debris (Figure 3.3).

The eruption 630,000 years ago has obscured much evidence of the two catastrophes that came before it, so our knowledge about the dimensions of those caldera explosions is uncertain. For that matter, the dates of all three eruptions also should be taken as the best current estimates.

The hole in the ground left by the 630,000-year-old eruption is the modern Yellowstone caldera, although it has since been mostly filled by smaller "post-caldera" lava flows. It is so large—and so obscured by subsequent volcanism, glaciation, and forest growth—that most park visitors are unaware they are inside a gigantic, active volcano.

Volcanism on a Grand Scale

The caldera explosions 2 million, 1.3 million, and 630,000 years ago were the "granddaddies of them all" in terms of volcanic eruptions at Yellowstone (Figures 3.4 and 3.5). The enormous scale of these eruptions—and the other huge caldera blasts that repaved the Snake River Plain between 16.5 and 4 million years ago—set the

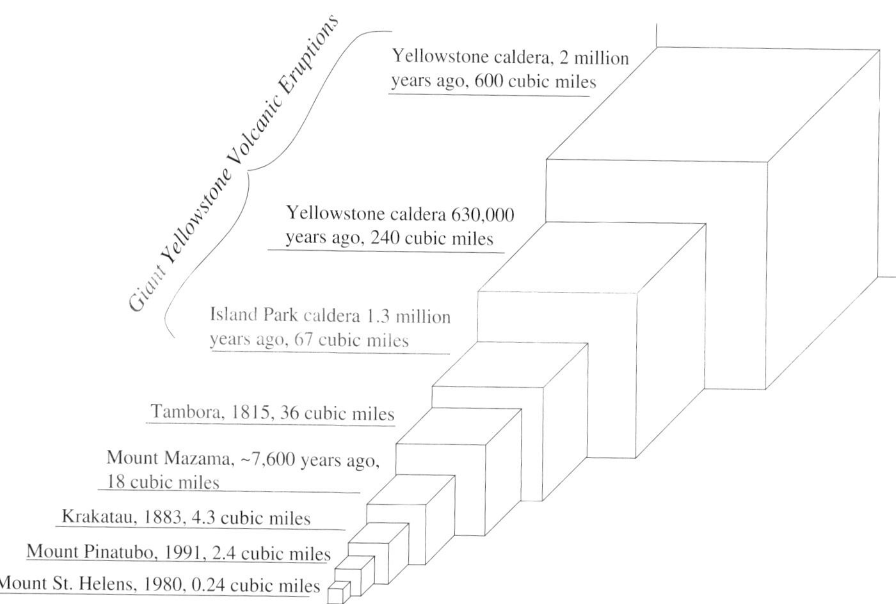

3.3 ~ *Volumes of Yellowstone's giant volcanic eruptions compared with volumes of other major eruptions.*

Yellowstone hotspot apart from the better-known but much smaller volcanic eruptions caused by hotspots beneath Hawaii and Iceland. The Yellowstone hotspot is the largest known center of active volcanism on any continent on Earth.

The three caldera explosions were not only much bigger than the Mount St. Helens eruption of 1980, they were larger than most prehistoric eruptions. Consider:

- The three Yellowstone caldera eruptions were 250, 28, and 100 times bigger than the 1991 eruption of Mount Pinatubo in the Philippines. Despite eruption predictions that saved thousands of lives, Pinatubo killed more than 700 people, forced closure of Clark Air Force Base—a U.S. military facility—expelled 2.4 cubic miles of ash, and slightly cooled Earth's climate for about two years.
- The amount of debris spewed by the Yellowstone eruptions was 33, 4, and 13 times larger, respectively, than the eruption of Mount Mazama in Oregon 7,600 years ago—a blast that left a large crater that now is the water-filled scenic heart of Crater Lake National Park.
- The Yellowstone eruptions were 140, 16, and 567 times larger than the infamous 1883 Krakatau volcanic explosion in what is now Indonesia, and 17, 2 and 7 times bigger than the gigantic 1815 eruption of Tambora, also in Indonesia.

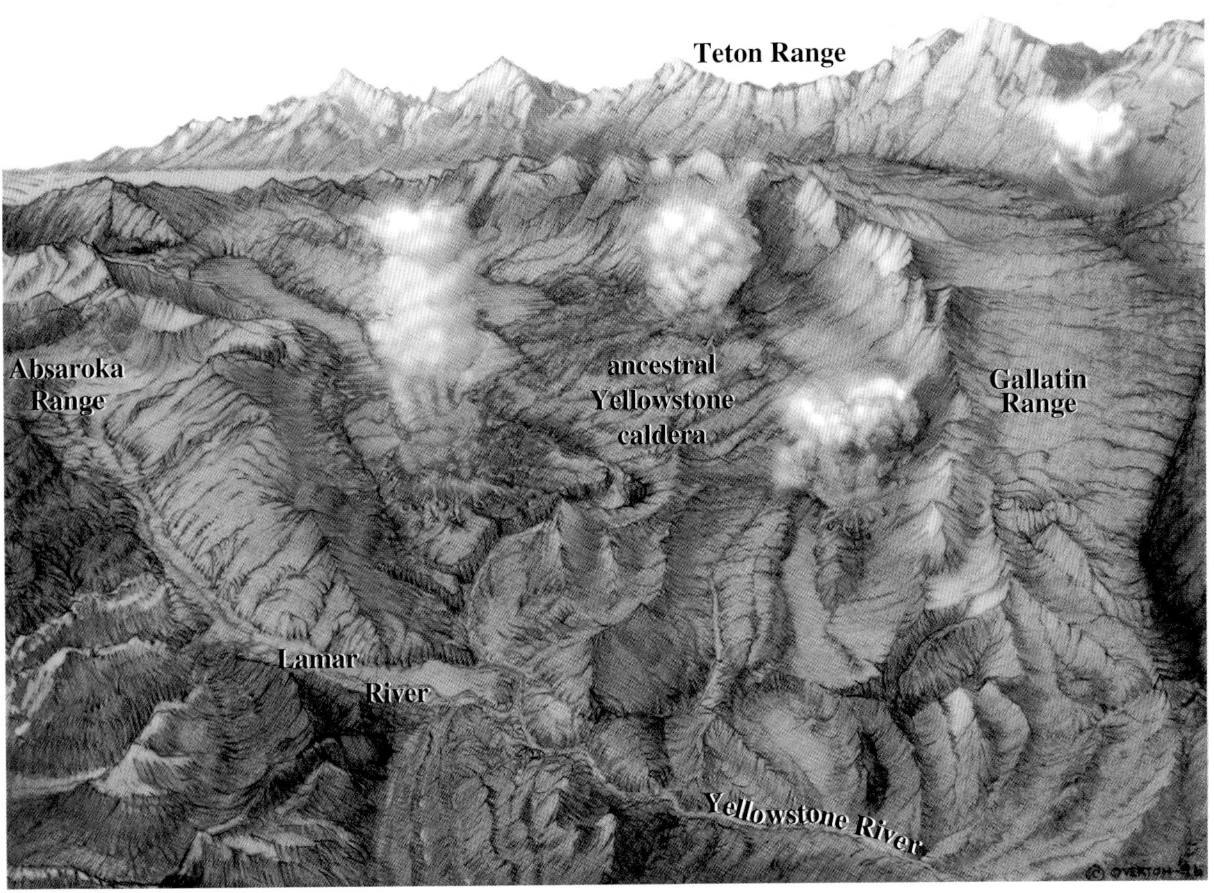

3.4 ~ *Artist's conception of Yellowstone's landscape, looking south, just before a giant caldera-forming eruption, showing how mountains once extended north–south across Yellowstone. Small eruptions develop first along buried faults and around the edge of the future caldera rim. (Chuck Overton.)*

Some prehistoric eruptions in North America were comparable in magnitude to the Yellowstone blasts. Mammoth Lakes, a ski resort on the east side of California's Sierra Nevada, sits inside the 10-by-20-mile Long Valley caldera, created by a massive eruption about 760,000 years ago. That explosion was about one-fourth the size of Yellowstone's largest caldera eruption but twice the size of the smallest caldera blast, the one at Island Park 1.3 million years ago. New Mexico's 14-mile-wide superimposed Valles and Toledo calderas erupted 1.2 and 1.6 million years ago, respectively. Each spewed an estimated 36 cubic miles of rock—making those eruptions half as big as the Yellowstone hotspot's Island Park caldera eruption.

3.5 ⁓ *Artist's conception of a cataclysmic caldera-forming eruption at Yellowstone spewing ash, gas, and molten rock. As the eruption progresses, the ground surface collapses to form a giant crater known as a caldera. (Chuck Overton.)*

The hot pumice and ash that fell to Earth near the Yellowstone and Long Valley caldera explosions formed welded tuffs, thick deposits of hot ash that solidified into hard rock. Tuffs from Yellowstone covered most of northwestern Wyoming, northeastern Idaho, and southern Montana.

Winds carried lighter ash great distances. The Yellowstone caldera blasts 2 million years ago and 630,000 years ago dropped volcanic ash as much as yards thick on most of what is now the western United States. The ash fell as far east as Minnesota, Iowa, and Missouri, as far south as Mexico, and, in the case of the 2-million-year-old eruption, as far west as a few hundred miles into the Pacific Ocean. Ash from the Long

Valley eruption in California went almost as far, extending southeast to Mexico and Texas; east into Oklahoma, Kansas, and Nebraska; and north into Oregon and Idaho. In contrast, ash from Mount St. Helens's 1980 eruption blew only about 700 miles away, falling on eastern Washington and parts of Oregon, Idaho, Montana, and Wyoming, including Yellowstone.

Together, Yellowstone's three gigantic eruptions spewed out 907 cubic miles of ash and rock. That's like a block of rock a mile high, a mile wide, and stretching 907 miles—more than the distance from New York to Atlanta.

Other eruptions, large but less catastrophic, also shaped Yellowstone. Between 2 million and 70,000 years ago, scores of smaller, but still explosive eruptions sent thick, slow-moving rhyolite lava flowing over the Yellowstone Plateau. The younger flows now cover much of the area west of Yellowstone Lake, including the Central, Madison, and Pitchstone plateaus. In addition, some basaltic lavas—the kind of flow seen in Hawaii—have erupted in Yellowstone as recently as 200,000 years ago.

The rhyolitic and basaltic flows produced almost as much lava as the 907 cubic miles of ash and debris hurled out by the caldera explosions: One conservative estimate suggests these smaller eruptions generated about 720 cubic miles of lava. The combined volume of ash, pumice, lava, and other debris totals at least 1,620 cubic miles—enough material to fill a cube measuring almost 12 miles on each side. That's more than enough to fill the Grand Canyon in Arizona. This is a conservative estimate because it includes only the three caldera blasts and lava flows after the latest one—but not the lava that flowed after the first two caldera eruptions and later was destroyed by the third caldera eruption.

The hotspot's 2 million years of volcanic activity at Yellowstone have been grouped into three cycles. Each cycle includes a caldera explosion, followed by extensive rhyolite lava flows. Now, 630,000 years after the last caldera cataclysm and 70,000 years after the most recent rhyolite flows, scientists do not know the significance of Yellowstone's current activity—its geysers and other hydrothermal features, earthquakes, upward and downward movements of the caldera floor, and the underground flow of molten rock or magma. Is such restlessness merely the waning stage of Yellowstone's third volcanic cycle? Or is it the beginning of a fourth cycle that eventually will climax in yet another unimaginable explosion? One fact remains certain: There is no evidence that volcanism has ceased at Yellowstone.

The history of Yellowstone's volcanic activity first came under scientific scrutiny with the pioneering expeditions of physician–geologist–naturalist Ferdinand V. Hayden in the 1870s. Trappers had hunted in Yellowstone earlier in the century, and other explorers visited during 1860 to 1870. However, it was during Hayden's 1871 expedition that a team of scientists mapped Yellowstone and sketched and photographed its geological and hydrothermal features. The vivid images struck the public imagination, leading to the designation of Yellowstone as the United States's first national park on March 1, 1872. Hayden conducted further explorations in 1872 and 1878. Scientific studies have continued with modern research by government and university geologists. From these studies, it is possible to describe when and how Yellowstone's calderas and smaller volcanoes formed, and how much of the scenery developed over time.

Rocks from Volcanic Ash

Hot ash from Yellowstone's three catastrophic explosions fell to the ground and hardened into rock called tuff, composed of fine-grained particles of rhyolite. The size of the eruptions is reflected by extensive tuff deposits found throughout Yellowstone and its surroundings.

The first catastrophic caldera eruption at Yellowstone produced the Huckleberry Ridge tuff, a 500 to 2,500-foot-thick sheet of volcanic rock. The caldera and the tuff are named for Huckleberry Ridge, a mountain along the southern boundary of Yellowstone park. The tuff layer is a brownish band of rock that can be seen by looking upward to Huckleberry Ridge itself, located east of Flagg Ranch, which is north of Grand Teton National Park and just south of Yellowstone's south entrance. The tuff is about 800 feet thick at the ridge. Other remnants of the Huckleberry Ridge tuff now are mostly covered by younger volcanic flows, but the rock formation is exposed on top of the northern end of the Teton Range; in Montana's Madison River Valley; and even more than 60 miles away near Idaho Falls, Idaho.

Yellowstone's second gargantuan blast, 1.3 million years ago, formed the Henry's Fork or Island Park caldera, which sits within the western end of the larger, 2-million-year-old caldera. Unlike the 2 million-year-old caldera, which is mostly obscured, the rim of Island Park caldera is quite evident to anyone driving from Ashton through Island Park, Idaho, and on to West Yellowstone, Montana. A rock layer named the Mesa Falls tuff was formed by hot ashfall from the Island Park caldera explosion. It is as much as 500 feet thick.

The third caldera blast came 630,000 years ago, centered more to the northeast. This tremendous explosion drained a huge magma chamber to create the most recent Yellowstone caldera and deposit the Lava Creek tuff, which in places is 1,600 feet thick. The modern topography of the Yellowstone Plateau *is* the flat floor of this

3.6 ~ *The modern Yellowstone landscape. After the last catastrophic caldera eruption, smaller lava flows filled the caldera and smoothed the terrain. Glaciers later gouged out lakes and valleys. The Yellowstone caldera occupies the central part of the park. The caldera floor is relatively flat and contains the western half of Yellowstone Lake, two resurgent lava domes, and most of Yellowstone's geysers. The Teton Range is seen in the distance. (Heinrich Beraan, with permission from the National Park Service.)*

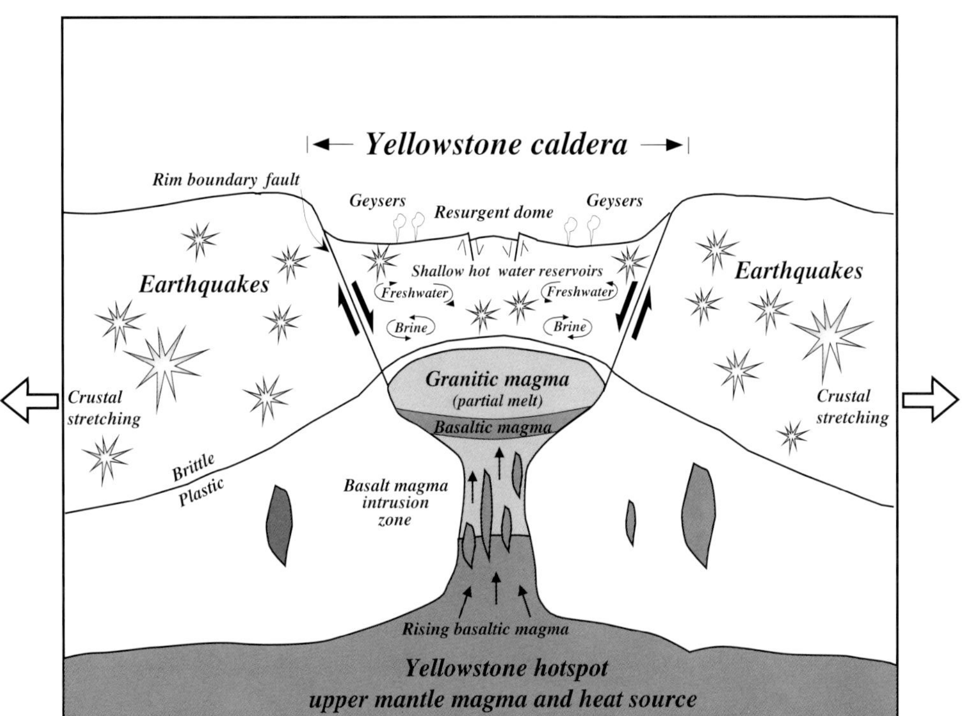

3.7 ~ *Volcanism, earthquakes, and geysers interact in the Yellowstone caldera. Heat and magma originate from the Yellowstone hotspot. Magma ascends into the overlying crust, melting surrounding rocks and lifting the ground surface. The molten rock explodes to the surface during eruptions. Then the ground collapses along faults to form a caldera. Pressures in the ground trigger earthquakes in cooler, brittle rock. The magma also heats groundwater from rain and snow to produce geysers and hot springs. (Adapted from Dzurisin, 1990.)*

caldera, altered by millennia of erosion, subsequent rhyolite lava flows, and glaciers (Figure 3.6).

This giant oval-shaped crater occupies essentially the entire central part of Yellowstone National Park. It extends about 45 miles from southwest to northeast and about 30 miles from southeast to northwest. Its western edge nears the national park's west boundary, where it overlaps the eastern portion of the 2-million-year-old caldera. The Yellowstone caldera's east side is just east of Yellowstone Lake.

During the eruption that formed the modern Yellowstone caldera, its floor collapsed more than 1,000 feet along a series of faults that form concentric rings inside and parallel to the caldera rim (Figure 3.7). As the caldera sank, lava started flowing out of the faults, or ring fractures. Then the whole caldera blew sky high. The slopes

3.8 ~ *The northwest rim of the Yellowstone caldera north of Madison Junction. The steep south-facing rim formed when the caldera floor* (foreground) *sank downward during the most recent giant eruption 630,000 years ago. Mount Holmes, a 10,336-foot peak* (background), *is part of the Gallatin Range; the southern end of the range was destroyed by Yellowstone's first caldera-forming eruption 2 million years ago.*

left when the caldera collapsed along the main ring fault can still be seen along the north side of the highway near Madison Junction. The north side of the canyon is the caldera rim (Figure 3.8).

Going with the Flow

Calderas are the most dramatic volcanoes in Yellowstone and the Snake River Plain, but they are not the only ones. Each of Yellowstone's three caldera explosions was followed, over hundreds of thousands of years, by smaller but still explosive eruptions of rhyolite lava similar in scale to Mount St. Helens's 1980 outburst.

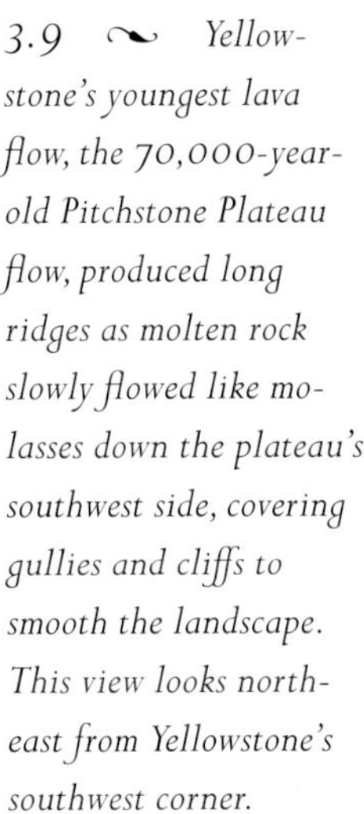

3.9 ~ *Yellowstone's youngest lava flow, the 70,000-year-old Pitchstone Plateau flow, produced long ridges as molten rock slowly flowed like molasses down the plateau's southwest side, covering gullies and cliffs to smooth the landscape. This view looks northeast from Yellowstone's southwest corner.*

These eruptions shot ash skyward and produced lava flows. The flows formed large cone-shaped hills or domes—a few hundred yards tall—of slowly moving, jagged, and jumbled rocks. A single flow covered from 10 to more than 100 square miles. The largest of these post-caldera eruptions produced more than 12 cubic miles of lava—about fifty times the volume of Mount St. Helens's 1980 eruption, or enough to fill a cube measuring 2.3 miles on each side.

Since the last caldera blowout at Yellowstone 630,000 years ago, thick rhyolite lavas have flowed about thirty times from volcanic vents in the caldera, most recently about 70,000 years ago on the southwest side of the caldera (Figure 3.9). These flows gradually filled in much but not all of the caldera, as if a baker had failed to apply frosting to the entire top of a cake. For example, a visitor at Old Faithful is standing in Upper Geyser Basin. The basin exists because it is surrounded by lava flows that stopped short of the area.

The rhyolite flows began in the northeast part of the caldera after the caldera explosion 630,000 years ago. Later eruptions moved generally southwest over hundreds of thousands of years, eventually covering and smoothing the pre-existing topography, and merging to form the gently rolling hills (Figure 3.10). Over time, erosion formed soils on top of the flows, which now are largely covered by lodgepole pine forests and broad meadows of the modern Yellowstone Plateau.

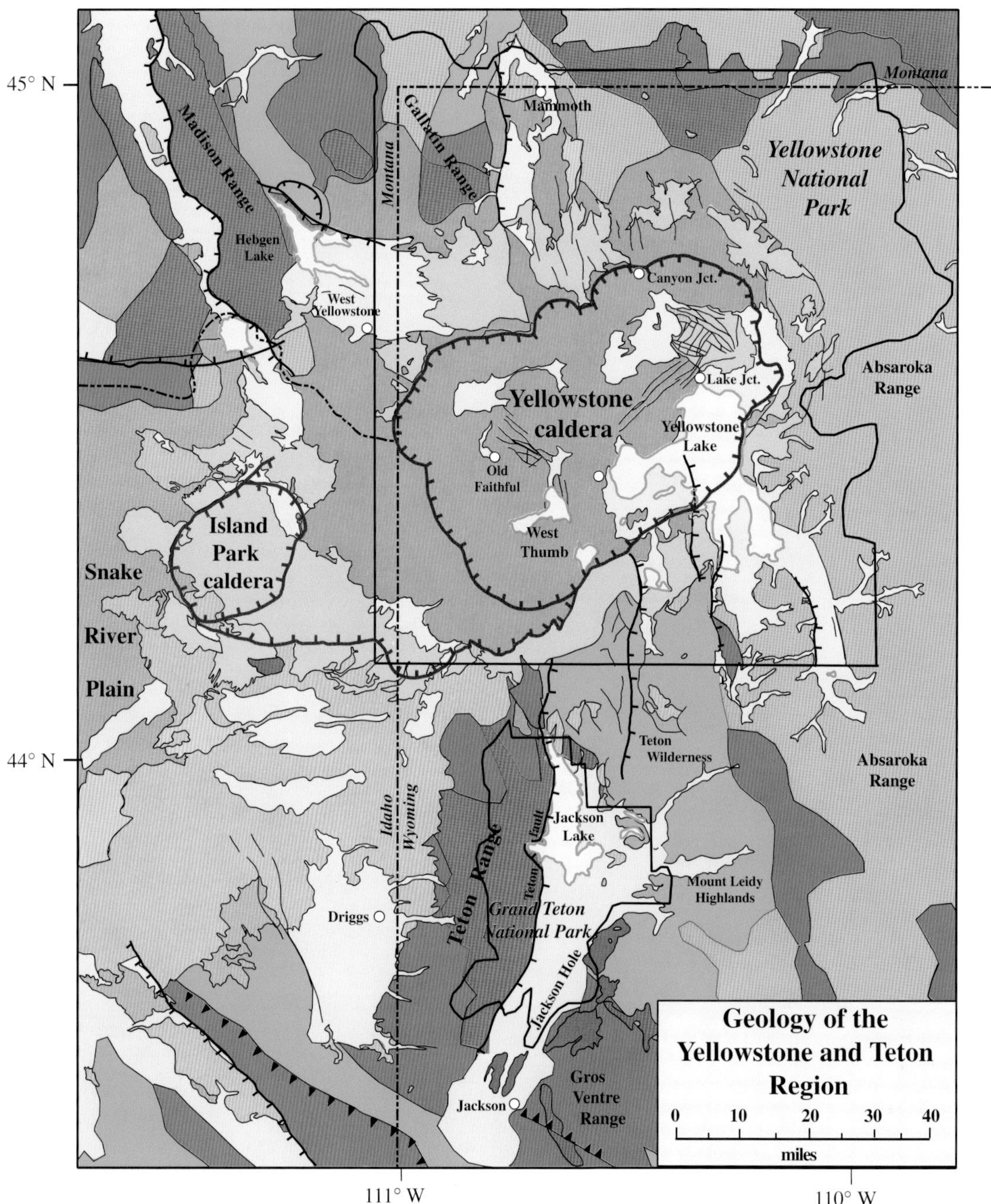

3.10 ～ *Lava flows that filled in much of the Yellowstone caldera are shown in this geologic map of the Yellowstone–Teton region. Rock units are colored by age and composition. Boundaries of the Yellowstone and Island Park calderas are hachured. (Adapted from Christiansen, in press; Smith, 1993; and U.S. Geological Survey maps.)*

The vents that produced the flows were lined up in two parallel zones stretching southeast to northwest. Both alignments coincide with pre-existing zones of weakness that cross the caldera—the same zones that are thought to be faults at the base of Teton-like mountains that once crossed Yellowstone.

Why are the vents and faults aligned from southeast to northwest? The vents and faults are essentially cracks in the Earth created because the crust is being stretched apart in a perpendicular, or southwest–northeast, direction. That stretching likely is caused by the same Basin and Range extension—the widening of the West by crustal movements—that has pulled apart much of Nevada, western Utah, and southern Idaho. In that region, the stretching produced the characteristic pattern of mountain ranges separated by broad valleys or basins. In Yellowstone, however, the typical basin-range-basin-range pattern was destroyed by caldera eruptions and then covered up by post-caldera rhyolite lava flows.

The flows were fed by underground magma chambers. The molten rock rose upward along the faults, forming vertical walls of molten rock called dikes. The dikes were scores of yards wide and tens of miles long, running southeast–northwest like the vents above them. The molten rock in the dikes erupted through the vents.

Along the Snake River Plain, caldera explosions and subsequent rhyolite lava flows were followed by eruptions of basaltic lavas. The basalts originated from deeper within the Earth. Only a few such basalt flows are seen in Yellowstone, mainly on the northern part of the Yellowstone Plateau.

South of Tower Junction near Calcite Springs, impressive vertical, hexagonal columns of basalt are exposed, looking somewhat like organ pipes. When the basalt flowed 1.3 million years ago, it formed a flat surface and then cooled, contracting into the columns. Each column is many inches wide. Columnar basalt is a common form of basalt. Another prominent basalt flow is located on Mount Everts, northeast and above Mammoth Hot Springs.

Legend

Rock Ages and Composition

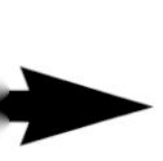

- 3.8 to 1.5 billion year old crystalline basement: granitic and metamorphic rock.
- 540 to 245 million year old limestone, sandstone, shale and dolomite.
- 245 to 65 million year old siltstone, limestone, sandstone and shale.
- 50 million year old volcanic breccia and andesite.
- 60 to 3 million year old volcanic conglomerate, tuff, andesite and claystone.
- 16 million to 10,000 year old basalt.
- First Yellowstone caldera, 2 million year old rhyolite flows and welded tuff.
- Second Yellowstone caldera, 1.3 million year old rhyolite flows and welded tuff.
- Third Yellowstone caldera, 630,000 year old rhyolite flows and welded tuff.
- 630,000 to 70,000 year old post-caldera rhyolite flows and ash fall.
- 1.6 million year old to present, glacial moraine and glacial outwash; gravel, sand and alluvium.

Volcanic Features and Faults

- 80 to 60 million year old thrust faults.
- 1.6 million years to present; normal faults, downdropped side is hachured.
- Secondary faults.
- Caldera, downdropped side hachured.

The Domes: Resurgent Yellowstone

During the past hundreds of thousands of years, two hills—called domes—have bulged upward from the Yellowstone caldera floor, and the entire floor has slowly huffed upward and puffed downward as hot water and magma move underground. The domes grew taller because of a resurgence of magma beneath the caldera. Both domes started as hills formed by erupting flows of rhyolite. Later, however, they bulged upward even more as magma pushed from below without erupting to the surface. The magma behaves like pistons, pushing up the domes without piercing the surface.

The Sour Creek dome rises about 1,000 feet above the surrounding countryside. The older of the two domes, it is located in the northeast part of the caldera, centered about 6 miles north of Fishing Bridge. It formed after the caldera catastrophe 630,000 years ago. It is an oval-shaped, pine-covered hill about 6 miles wide and some 10 miles long from southeast to northwest.

The Mallard Lake dome, a mile north of Old Faithful in the southwest half of the caldera, started bulging upward about 150,000 years ago. It also is slightly oval-shaped, trending southeast–northwest. It is about 7 miles long and 5 miles wide (Figure 3.11).

When a loaf of bread bakes and expands, a broad crack often forms on top of it, extending from near one end to the other. In a similar manner, the top of each dome is cracked open from northwest to southeast by a small valley a half mile to 1.2 miles wide and perhaps 6 miles long. Each valley dropped down along two parallel faults, which also extend northwest–southeast along the top of each dome. This kind of small valley, called a graben, forms when a block of ground drops down between the two faults. The presence of a graben on each dome suggests the domes expanded in volume as they bulged upward and outward.

The oblong shape of 45-mile-long Yellowstone caldera and the presence of two domes within the caldera—one to the northeast and one to the southwest—suggests that a magma chamber in the crust feeds magma upward through two pipe-like conduits, which emerge at the surface about 20 miles apart. These conduits fed Yellowstone's volcanism during the caldera blast 630,000 years ago, and have continued to feed the domes ever since, making them rise and fall. The dual conduits give the caldera its elongated shape.

Elephant Back Mountain, a ridge that exceeds 8,600 feet in elevation, extends 6 miles from southwest to northeast between the two domes but does not connect them. Northeast-trending faults extend along the length of Elephant Back Mountain. These faults probably accommodated the upward bulging of Elephant Back, as well as strains

3.11 ⁓ *The Mallard Lake resurgent dome is in the center, between Upper Geyser Basin* (foreground) *and frozen Yellowstone Lake* (background). *The top of the dome was lifted 1,200 feet above the surrounding terrain by molten rock beneath Yellowstone. (Rick Hutchinson.)*

that stretched the mountain from northwest to southeast. It is not unreasonable to assume that the same magma conduits pushing up the two domes also helped push up Elephant Back Mountain. Measurements of how Earth's crust is deforming at Yellowstone suggest that a connection between the two domes helps to evenly distribute the pressure of rising magma and hot water, making the entire caldera floor rise or fall as a unit.

Hot Water and Steam Eruptions

Some of Yellowstone's smallest eruptions are also explosive. These blasts do not expel pulverized molten rock. Instead, they are explosions of hot water and steam known as hydrothermal eruptions or phreatic eruptions, from the Greek word for "well." They are much bigger and more violent than geysers.

Hydrothermal eruptions happen when hot, solid rock about a mile deep heats up deep groundwater that is briny due to chemical reactions with surrounding rock. The deep brine, in turn, heats large volumes of shallower, fresh groundwater. Pressures build underground, and the superheated water flashes into steam, exploding violently to blow out craters a few hundred yards to a few miles wide. During the Ice Age, the weight of the glaciers acted like a pressure-cooker lid, allowing even greater pressures to accumulate underground. That pressure was released after the glaciers receded about 14,000 years ago—and also during earlier interglacial warm spells—resulting in particularly explosive steam eruptions.

Hydrothermal eruptions break overlying surface rocks into small fragments called breccias, which hurtle outward in all directions, forming a rim 10 to 30 yards high encircling the explosion crater. Phreatic craters are well known around the east side of the Yellowstone Plateau. They include Indian Pond, Mary Bay, and other small lakes and bays on the north end of Yellowstone Lake. The shattered rocks from the Mary Bay explosion can be seen in outcrops on the northeast side of Yellowstone Lake. Pocket Basin, located in Lower Geyser Basin about 8 miles north of Old Faithful, was formed by hydrothermal explosions that produced extensive deposits of shattered rock in the area.

The relatively small scale of phreatic eruptions means that older ones have been covered or obliterated by more recent volcanic or nonvolcanic rocks. Yellowstone's oldest known hydrothermal explosion craters are thousands of years old, but such steam blasts likely punctuated the entire 2 million years of hotspot volcanism at Yellowstone.

All the geysers, hot springs, steam vents, and earthquakes at Yellowstone today are only minor reminders of the last 2 million violent years of cataclysmic caldera blasts, extensive rhyolite flows, and explosive steam eruptions.

How Yellowstone Works

4

Most people who visit Yellowstone are blissfully unaware they are standing on top of an active, breathing volcano. They visit geysers and hot springs, and may feel some of the numerous earthquakes that rattle the region. Few realize the seemingly solid ground beneath them is slowly stretching apart and huffing and puffing upward and downward.

Nor are many visitors aware of the large chamber of molten and partially molten rock several miles beneath their feet, or of the even deeper plume of hot rock moving up from deep within Earth. Indeed, it is easy to enjoy the national park's geysers and other scenery without stopping to consider they are merely the uppermost, most visible parts of one of the world's geological wonders: the Yellowstone hotspot.

Even fewer tourists realize the same forces driving Yellowstone's renowned geysers also reshaped the landscape of 25 percent of the northwestern United States—a broad band stretching from Yellowstone almost 500 miles southwest to the Idaho–Oregon–Nevada border. As North America drifted southwest over the hotspot during the past 16.5 million years, the immense heat and molten rock rising from Earth's mantle melted, rearranged, and blew apart the overlying crust.

Today, the hotspot is beneath Yellowstone, making the national park a field laboratory of active geologic process: volcanism, earthquakes, faulting, and large-scale move-

4.1 ∾ Yellowstone's most famous geyser, Old Faithful, Upper Geyser Basin, erupts about every 80 minutes to heights of 100 to 180 feet. Each eruption releases up to 8,500 gallons of hot water. (Rick Hutchinson.)

ment and deformation of Earth's crust. Let us examine how this system works—how heat and magma, or molten rock, from within the Earth drive small-scale features such as geysers and hot springs, contribute to the most intense earthquake and volcanic activity in the Rocky Mountains, and help mold the topography of the region.

Heat Is Yellowstone's Driving Force

The amount of heat flowing from the ground in the Yellowstone caldera is thirty to forty times more than the heat emitted by an average piece of ground elsewhere on Earth's continents.

This enormous heat flow provides the energy that melted rock under the caldera and helped lift Yellowstone to its lofty altitude. Heat powers Yellowstone's volcanic activity by melting rock in Earth's mantle and crust. In turn, the molten rock heats groundwater to produce geysers and hot springs. Heat influences where earthquakes happen in Yellowstone, the Snake River Plain, and adjacent highlands. Heat is also the ultimate force that drives the movements of Earth's crustal plates and makes North America drift over the hotspot.

Heat in the Earth is released in two ways:

1. Conduction, or the movement of heat from hotter to colder rock. A common example of conduction is when heat from a stove is transferred through the bottom of a coffee pot to the liquid inside. Conduction in the Yellowstone hotspot helps transfer heat from deep within Earth to shallower depths. Of heat released from the ground at Yellowstone, about 25 percent comes from conduction.

2. Convection, in which heat is transported by flowing hot fluids such as hot water or magma. Convection happens inside a coffee pot when heat is carried upward by hot water that rises buoyantly because it is less dense than cooler water. As the water boils, the rise and fall of water forms what is called a convection cell. Convection of molten rock also helps carry heat up through the Yellowstone hotspot. Near the surface, convection of hot groundwater drives geysers, hot springs, and fumaroles or steam vents. Convection accounts for roughly 75 percent of the heat released from the ground at Yellowstone.

Despite Yellowstone's extraordinary hydrothermal attractions, few direct measurements have been made of the amount of heat flowing from the ground in the park. Boreholes have been drilled a few hundred to almost 1,500 feet deep to measure groundwater chemistry, rock composition, and ground temperatures. The hottest borehole temperatures were about 460 degrees Fahrenheit. This shows that rocks and groundwater are unusually hot close to the Earth's surface at Yellowstone. Researchers also have dropped temperature probes from a boat on Yellowstone Lake to measure the amount of heat flowing upward through lake-bottom mud. Much of the lake sits inside the Yellowstone caldera. The lake's southern arms are outside the caldera. Not surprisingly, the amount of heat flowing through the lake bottom was five to thirteen times greater inside the caldera than outside it. This dramatic difference provides more compelling evidence that beneath the caldera, there is widespread hot groundwater and, at greater depths, molten and partially molten rock.

On average at Yellowstone, a square meter of ground—somewhat more than a square yard—emits an amount of heat equal to 2 watts. So if the heat released from 50 square meters of ground was converted to electricity, it would illuminate a 100-watt light bulb. Yellowstone as a whole emits about 5 gigawatts of energy—enough to power a city of more than 2 million people.

Yellowstone's heat output not only vastly exceeds heat emitted by average continental rock, it also is twenty times more than the amount of heat released from relatively warm areas of Earth's crust such as the Snake River Plain or the Basin and Range Province. Both are being stretched apart, thinning Earth's crust so heat escapes more easily and the crust is warmer than in thicker, colder rock.

In addition to producing Yellowstone's well-known hydrothermal attractions, the high heat flow created a natural, underground oil refinery in the northern part of the park. Heating of sediments rich in organic materials has made gas and oil seep to the surface in places such as Calcite Springs and Rainbow Hot Springs. Elk and bison roll in oily soil at Rainbow during the summer when insects are particularly pesky. Yet exploratory drilling failed to find significant oil fields around Yellowstone. Some geologists theorize high heat flow has cooked many hydrocarbons out of the ground.

What does it take to produce the heat observed at Yellowstone? How much new molten rock must move upward from the hotspot and melt shallower rocks, which then cool and solidify beneath the park to produce the observed flow of heat?

One estimate is that each year, the amount of new molten rock beneath Yellowstone would equal a cube measuring about 900 feet on each side—as long as three football fields end-to-end—and the magma would have to cool from 1,580 to 840 degrees Fahrenheit. Of course, the magma injected beneath Yellowstone more likely is shaped like a huge pie beneath the caldera. The large amount of heat flowing from the Earth at Yellowstone is strong evidence new magma is emplaced continuously beneath the park, providing the energy to sustain geysers, hot springs, and other hydrothermal features.

Hot Rocks in Earth's Crust

Other measurements reveal more about Yellowstone's underground geology. For example, rocks beneath Yellowstone are less dense than rocks in nonvolcanic settings because high temperatures expand their volume and they are highly fractured. That allows steam, hot water, and molten rock to intrude into cracks and chemically alter the composition of the rocks. When a geographic area sits atop rocks that are lighter, or less dense, the pull of gravity in that area is weaker than in areas of denser rocks.

Measurements show the pull of Earth's gravity is 30 percent below normal on the Yellowstone Plateau and in an area extending about 15 miles northeast. That suggests hot water and hot and molten rocks not only are beneath Yellowstone, but extend outside the caldera to the northeast—not surprising because North America continues to slide slowly southwest over the hotspot.

A more detailed picture of what lies beneath Yellowstone comes from measuring how fast seismic waves from earthquakes or sound waves from explosions move through the Earth, then using that information to construct cross-sectional images of underground geology. Seismic waves travel faster through cold, dense rock and more slowly through hot or molten rock, which is less dense.

Seismic images reveal Yellowstone's subsurface rocks include a 3,000- to 6,000-foot-thick top layer made up of solidified rhyolite lava flows capped by more recent sediments. Beneath that, extending as deep as 8 miles, is a giant pluton—a blob of granite that is hot but mostly not molten. However, within the pluton is the magma chamber.

Many people envisage Yellowstone's magma chamber as an underground lake of molten rock within which one could row a fireproof boat. However, seismic images indicate the molten rock beneath Yellowstone is contained in interfingering passageways within a sponge-like volume of solid rock. This zone of spongy, partly molten rock begins about 5 miles beneath most of the caldera and extends down to 8 miles. Beneath the caldera's northeast corner, some partly molten rock may be as shallow as 3 miles deep. This partly molten rock is only about 10 to 30 percent liquid magma in a matrix of hot but solid rock (Figure 4.2).

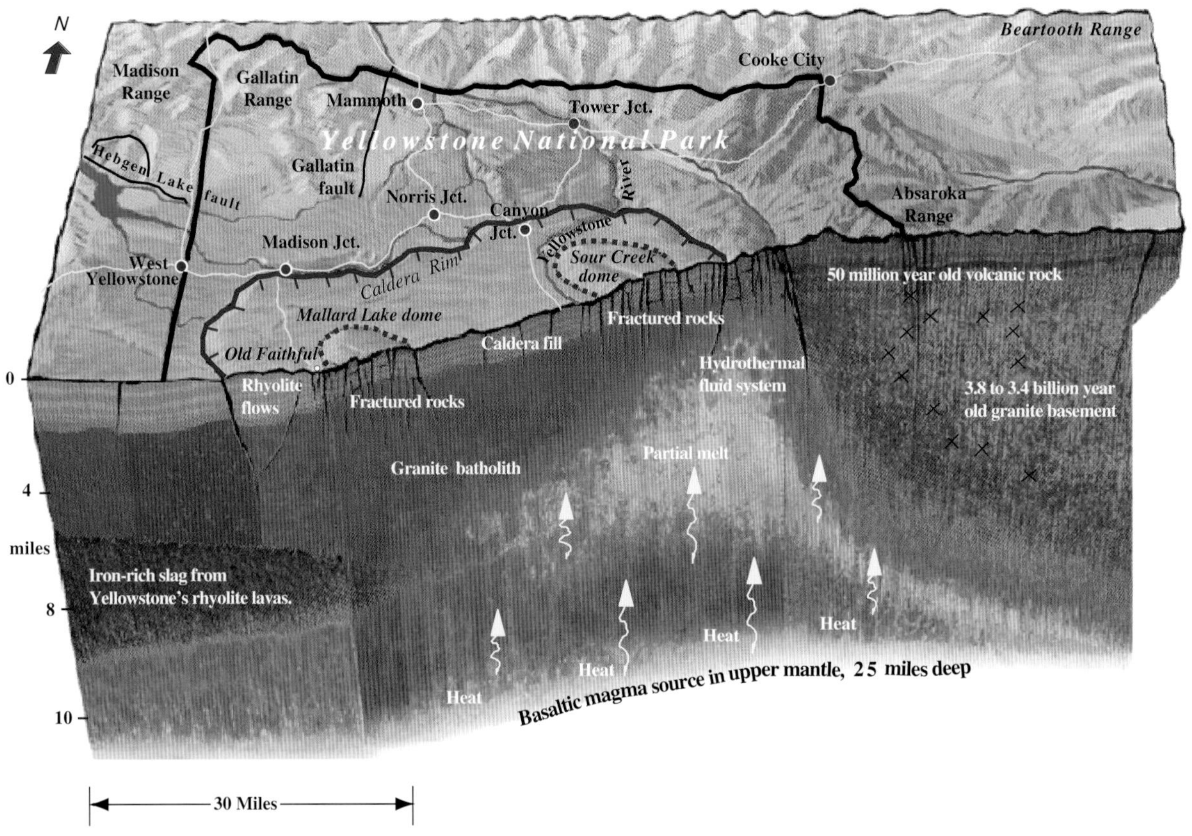

4.2 *A cross section of Yellowstone reveals molten rock under the caldera at depths of about 3 to 8 miles. Heat emitted by the molten rock powers Yellowstone's geysers and hot springs.*

At greater depths, in the lower crust, there must be conduits that allow magma to move upward from Earth's mantle. Rocks surrounding the hotspot appear similar to the colder rocks in the rest of the Rocky Mountains.

Yellowstone's Geysers and Hot Springs

Prehistoric Indian relics have been found at Mammoth Hot Springs and in Yellowstone's geyser basins (Figure 4.3), and accounts tell of Native Americans bathing in geysers and using hot-springs water for cooking. Yet Yellowstone's geysers also inspired fear among some early Indians. They interpreted geyser eruptions as "the result of combat between the infernal spirits," according to Father Pierre-Jean De Smet,

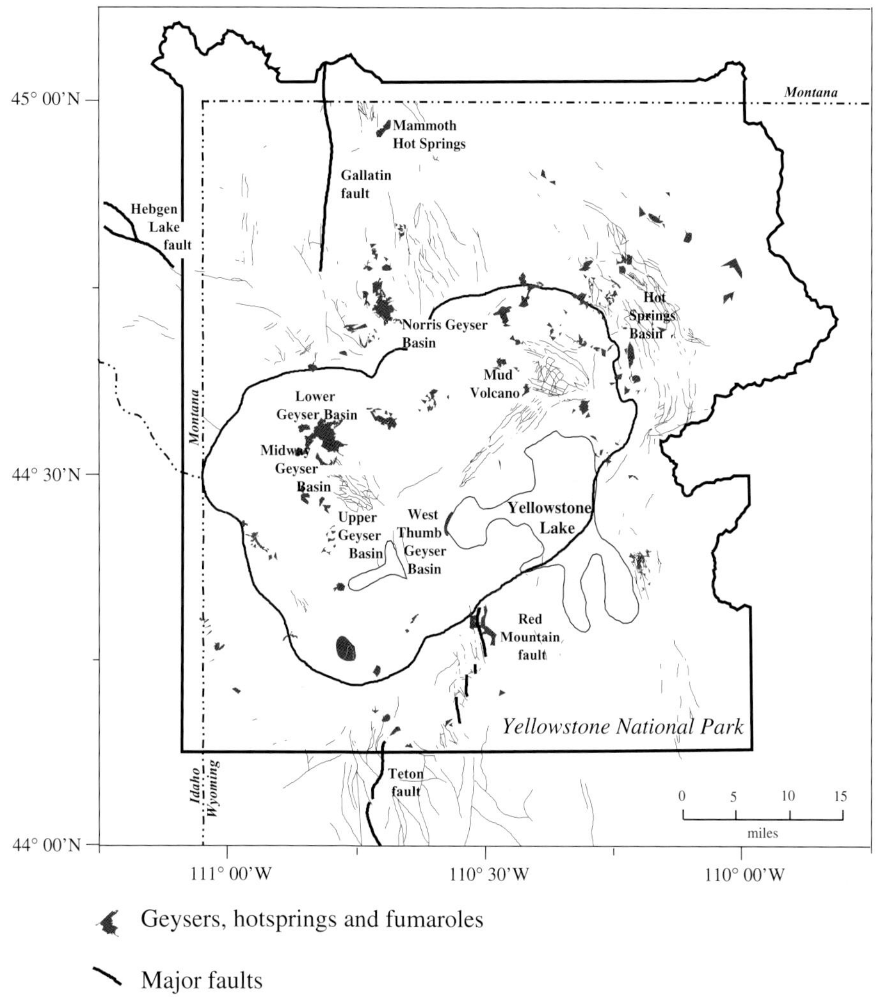

4.3 ⁓ *Map of Yellowstone's faults and hydrothermal features (geysers, hot springs, and steam fumaroles). (Adapted from Rick Hutchinson unpublished data and Christiansen, in press.)*

a nineteenth-century priest. The famed expedition of Meriwether Lewis and William Clark traveled north of Yellowstone in 1806. Clark later told of the Indians' concern: "Very frequently there is a loud noise heard like thunder which makes the earth tremble—they state they seldom go there because their children cannot sleep at night for this noise and conceive it possessed of spirits who are averse that men should be near them."

The first white man known to have traveled through Yellowstone was John Colter, who left the Lewis and Clark expedition and observed what he called "hot spring

4.4 ⁓ An aerial view of a rare eruption of Steamboat Geyser, Norris Geyser Basin. Steamboat Geyser is the world's tallest geyser but erupts very irregularly at intervals of four days to fifty years. It sends water 390 feet high and steam another 350 feet or more. (Robert B. Smith, from an extraordinarily lucky photo opportunity, July 6, 1984.)

brimstone" while crossing Yellowstone. To beaver trapper Joseph Meek in 1830, Yellowstone's hydrothermal attractions inspired thoughts of smoke rising from industrial Pittsburgh on a winter morning—and "about hell and the day of doom," according to one account. William Ferris, another trapper in the same era, reported: "The continual roaring of the springs . . . for some time prevented me from my going to sleep. . . . Clouds of vapor seemed like a dense fog to overhang the springs, from which frequent reports of explosions . . . constantly assailed our ears." Trapper Jim Bridger, whose tall tales often overshadowed his accurate observations about Yellowstone in the 1850s, called it "the place where Hell bubbled up."

Yellowstone is renowned for the world's largest and most spectacular display of geysers, hot springs, and steam vents (also called fumaroles) (Figure 4.4). Geysers were named after the Icelandic word *geysir*, which means gushing. Some 200 to 250 geysers spout off each year at Yellowstone—a greater concentration of geysers than anywhere else in the world. The park also has thousands of hot springs and steam vents.

These hydrothermal features are located in places where rainwater and snowmelt can sink easily into the ground, become superheated by underlying magma, then easily erupt to the surface again. That is why so many of the geysers and hot springs are in flat-bottomed valleys between lava flows. Such flows erupted after the last major caldera eruption about 630,000 years ago. The lava covers much of the Yellowstone Plateau like frosting on top of a cake, but there are places where the flows did not converge. These valleys are covered by permeable stream and glacier sediments that allow water to percolate into the ground where it can be heated.

Yellowstone's geysers also tend to be located in other places where water can penetrate easily underground, including along active faults inside and outside the caldera, and at the bottom of slopes where groundwater pools after running off higher ground. Geysers also are found on the ring-shaped fracture zone along which the Yellowstone caldera floor sank downward during its last catastrophic eruption.

A key theory of how geysers and hot springs work involves the transfer of heat from brine, which is salty water, 1.5 to 3 miles deep. According to this concept, hot or molten rock at depth heats the overlying brine (Figure 4.5). Like water boiling in a pot, the hot brine carries heat upward by convection. That, in turn, heats overlying, fresh groundwater circulating closer to the surface. The process is aided by the presence of highly fractured, porous rocks that allow groundwater to flow easily.

When it is confined, deeper groundwater becomes superheated—exceeds the normal boiling point—and pressurized, like the contents of an old-fashioned pressure cooker. When enough pressure builds up, the superheated water overcomes the

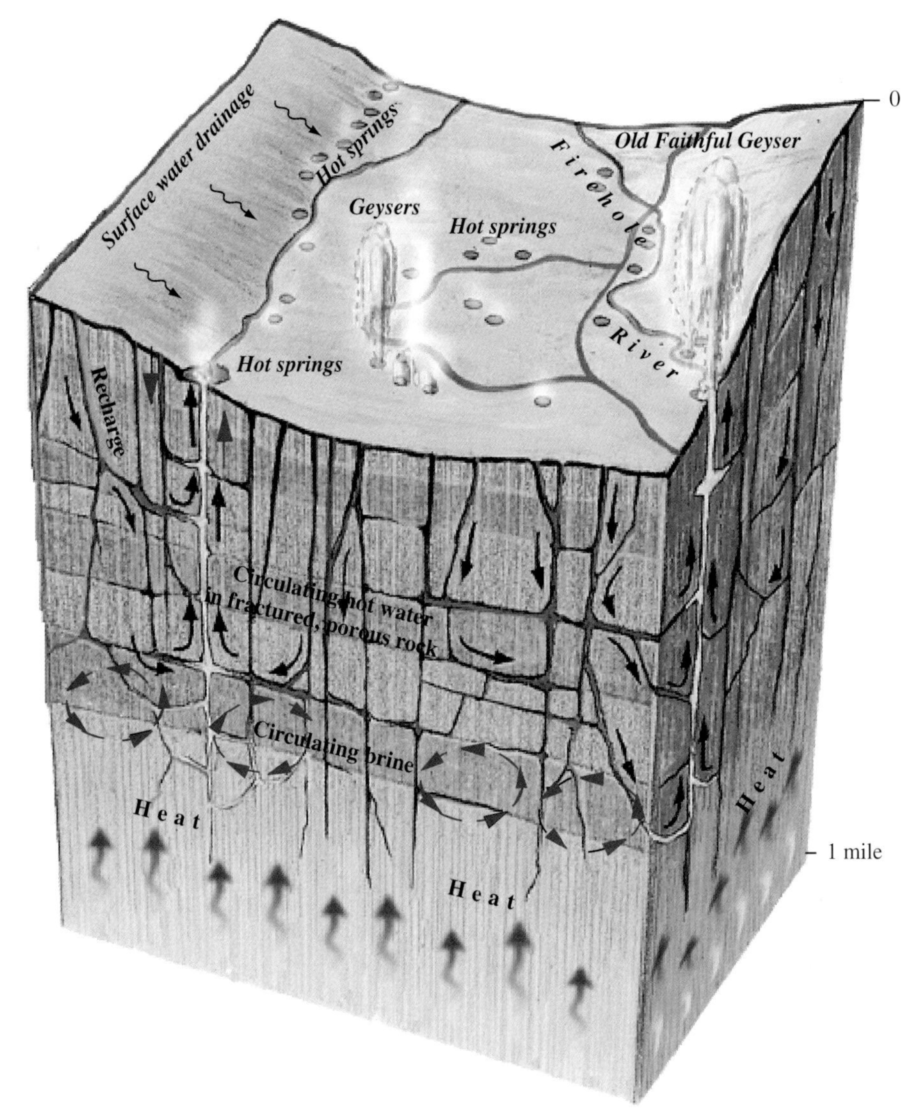

4.5 ~ *Workings of a Yellowstone geyser illustrate how heat from cooling magma rises to warm dense salty water circulating in a porous rock layer. The brine transfers heat to overlying fresh groundwater, which is recharged by rainfall and snowmelt. The shallow hot water can be trapped and pressurized in small underground reservoirs, then released suddenly through narrow openings, creating geysers.*

weight and pressure of cooler groundwater above it, allowing steam and hot water to burst out of a geyser.

Once a geyser starts erupting, it is self-perpetuating. The initial eruption of hot water reduces pressure in the superheated groundwater, which then flashes to steam. The steam blasts upward, gaining force as it passes through a narrow opening, then hurling out any overlying water. The process is quite similar to what happens when a pressure cooker's relief valve suddenly opens: Hot water and steam are expelled at high speeds through a narrow opening. Even in a large geyser, the steam and water erupt through an underground conduit that is relatively small. The opening through which water spurts to form Old Faithful is a slot or crack measuring only about 2 by 5 feet at the surface. However, the real action occurs about 22 feet underground. There, superheated water and steam surge upward at supersonic speeds during eruptions as they pass through a slot only about 4 inches wide by perhaps a few feet in width. During each of the geyser's 2- to 5-minute eruptions, 3,700 to 8,500 gallons of boiling water spurt through that conduit. Water inside Old Faithful has been measured up to 244 degrees Fahrenheit—well above boiling—and steam temperatures reach 265 degrees Fahrenheit.

Hot springs occur when rising hot water is not trapped in a pressurized reservoir, but simply flows to the surface. If relatively little hot water flows through clays or other colored soils, chemicals in the water break down the clays to form bright red, yellow, orange, gray, or even black mudpots or "paint pots" that burp and belch steaming gases, some with the distinctive rotten-egg smell of hydrogen sulfide. Examples include Mud Volcano and Black Dragon's Caldron in Hayden Valley, and Fountain Paint Pots in Lower Geyser Basin.

When only steam is released without much hot water, the hydrothermal feature is a fumarole. Such steam vents often are encrusted with deposits of minerals that had been dissolved in the vent's hot water and steam supply. Some fumaroles expel steam continuously at such high speed that they sound like jet engines and sometimes can be heard miles away.

Hot Springs Basin, a relatively little-known and remote area 15 miles north–northeast of Fishing Bridge, has one of Yellowstone's largest areas of hot springs and fumaroles (Figure 4.6). The fumaroles produce mounds of sulfur a few feet high. Water in Hot Springs Basin is so acidic that it has dissolved the pants of people who sat on moist ground. The ground sounds hollow in places where the acidic water dissolved rocks to create voids just inches beneath the surface. These voids contain steam and hot water. People walking across Hot Springs Basin and other hydrothermal basins in Yellowstone have been burned seriously after breaking through a thin crust into the hot water.

Chemical reactions from heat and hot water also changed the color of rhyolite lava deposits to create the vivid yellow to pink colors exposed in the walls of the Grand Canyon of the Yellowstone River. Contrary to popular belief, Yellowstone was not named for these rocks. Instead, the name was cited as early as 1805 by Indians who were referring to yellow sandstones along the banks of the Yellowstone River in eastern Montana, several hundred miles downstream and northeast of the national park.

Geysers and hot springs usually cover surrounding ground with deposits of geyserite or sinter—a white to gray, grainy mineral made of silica and opal, which is quartz with extra water in it (Figure 4.7). Sinter deposits occur where hot water percolates upward through silica-rich bedrock, usually solidified rhyolite lava inside the Yellowstone caldera. After thousands of years, sinter deposits built layer upon layer, creating cones, mounds, and other unusual features that inspired nineteenth-century explorers to name geysers such as Castle, Cone, Grotto, and Dome.

An even whiter mineral—a form of calcium carbonate or limestone named travertine—is common in hot springs outside the caldera, notably in the terraces at Mammoth Hot Springs. Hot, acidic water dissolves underground limestone, which was deposited on an ancient seafloor. When the water surfaces and loses acidity, the limestone precipitates as travertine. Other minerals in Yellowstone's hot springs include

4.6 ~ *The Hot Springs Basin Group of springs and steam vents, 15 miles north–northeast of Fishing Bridge, is among the largest thermal areas in Yellowstone. The basin has yellowish-gray mounds of sulfur and colored mineral deposits. (Charles Meertens.)*

iron oxides, which contribute black and gray colors to Mud Volcano, and bright yellow sulfur mounds at Hot Springs Basin.

The colors seen in hot springs can come from sources other than minerals. Light hitting particles in the water give a delicate blue hue to Grand Prismatic Spring and to Morning Glory, Opal, and Sapphire pools. Algae and other microbes can flourish at temperatures exceeding 170 degrees Fahrenheit, growing in thick mats that add bright green, yellow, and orange colors to some hot springs (Figure 4.8).

Microorganisms growing in Yellowstone's hot springs have found their way into modern science and criminal law. In the mid-1960s, biologist Thomas Brock discovered a new microbe—which he named *Thermus aquaticus*—in hot water from Mushroom Pool. Two decades later, Kary Mullis and colleagues at Cetus Corp. were able to harness Taq—an enzyme produced by *Thermus aquaticus*—in a process named polymerase chain reaction, or PCR. This method is able to quickly produce millions of copies of DNA, the genetic material found in every living cell. Mullis earned the 1993 Nobel Prize in chemistry for PCR, which helps doctors to diagnose infectious diseases and scientists to identify disease-causing genes. PCR also is used by police and prosecutors to identify criminals by making millions of copies of the DNA found in hair, blood, or other body fluids and tissues. This method was used in O. J. Simpson's

4.7 ~ *Mineral deposits called geyserite dam a pool of hot water* (foreground) *and form a 3-foot-high mound* (background) *called the Sphinx at Rainbow Hot Springs, just north of Hot Springs Basin. (Robert B. Smith.)*

murder trial, during which the defense called Mullis to testify on PCR's shortcomings. The extraction and commercialization of Yellowstone's hot-water microbes have triggered debate over how much the government should be paid by so-called bioprospectors.

Yellowstone's most prominent geysers and hot springs are found in six areas: the Lower, Midway, and Upper geyser basins—including Old Faithful—on the southwest side of the caldera; Norris Geyser Basin's Porcelain Basin, where geyserite has formed a porcelain-white landscape; Mud Volcano in Hayden Valley; and Mammoth Hot Springs.

Some of Yellowstone's geysers erupt continuously. Old Faithful and many other geysers are famous for spouting off at intervals ranging from minutes to hours, days, months, and even years. The timing of geyser eruptions depends on the amount of

4.8 ~ *The edges of Emerald Pool, a hot spring in Upper Geyser Basin, are colored by heat-resistant microbes. (Rick Hutchinson.)*

heat and water that powers them. As groundwater is heated from below, it must reach a critical temperature and pressure before it erupts to the surface. How often it reaches that temperature depends, in turn, on the initial temperature, volume, and speed of water flowing into the underground hydrothermal system. The extent to which rocks are fractured, allowing groundwater to flow, is also a factor.

While Old Faithful is Yellowstone's most famous geyser, it is not the tallest, erupting to heights of 100 to 180 feet. Steamboat Geyser, which rarely erupts, is the world's tallest, rising 230 to 390 feet above Norris Geyser Basin. Giant Geyser erupts as much as 200 feet above Upper Geyser Basin. Hours to years can elapse between eruptions of some of the taller geysers, making Old Faithful the most reliable of the big geysers because it spouts off roughly every 80 minutes, although that interval was once shorter.

Earthquakes in Yellowstone

During the summer of 1871, explorer Ferdinand Hayden stood on the shore of Yellowstone Lake during a pioneering field expedition. He later wrote in his journals:

> While we were encamped on the northeast side of the lake on the night of the 20th of July, we experienced several severe shocks of an earthquake, and these were felt by two other parties, fifteen to twenty-five miles distant, on different sides of the lake. We were informed by mountain-men that these earthquake shocks are not uncommon, and at some seasons of the year very severe. . . . I have no doubt that if this part of the country should ever be settled and careful observations made, it will be found that earthquake shocks are of very common occurrence.

The tremors were so persistent that Hayden named the site Earthquake Camp. More than a century after his visit, Yellowstone Lake remains one of the region's most notable sites for so-called swarms of quakes.

Earthquakes are manifestations of active faulting, mountain-building, and, in some cases, volcanic activity. They are prominent indications that these processes are active in Yellowstone. The area produces a disproportionate share of earthquakes in the 800-mile-long, 120-mile-wide Intermountain Seismic Belt, a band of frequent earthquakes that stretches from southern Nevada north through Utah, straddles the Idaho–Wyoming border and continues north into Montana. The Yellowstone Plateau represents about 1 percent of the total area in the Intermountain Seismic Belt, yet it generates up to 20 percent of its earthquake energy. The Yellowstone area is the most intense region of seismicity in the Rocky Mountains (Figure 4.9).

The 1959 magnitude-7.5 Hebgen Lake earthquake, centered near the lake on the west side of the Yellowstone Plateau, was the largest quake in the Rockies in historic time. It killed twenty-eight campers, most of whom were buried or drowned when shaking triggered the huge landslide that dammed the Madison River. The same kind of quake would spell catastrophe for a densely populated urban area like Salt Lake City, Utah, which sits on top of a similar fault zone.

Earthquakes in Yellowstone have been studied extensively only since the Hebgen Lake disaster. Because of its remote location, Yellowstone did not have permanent seismographs until the early 1970s, when such devices were installed by the U. S. Geological Survey. Starting in the early 1980s, the University of Utah began operating modern seismographs that record data on quakes, then use microwave facilities and satellites to

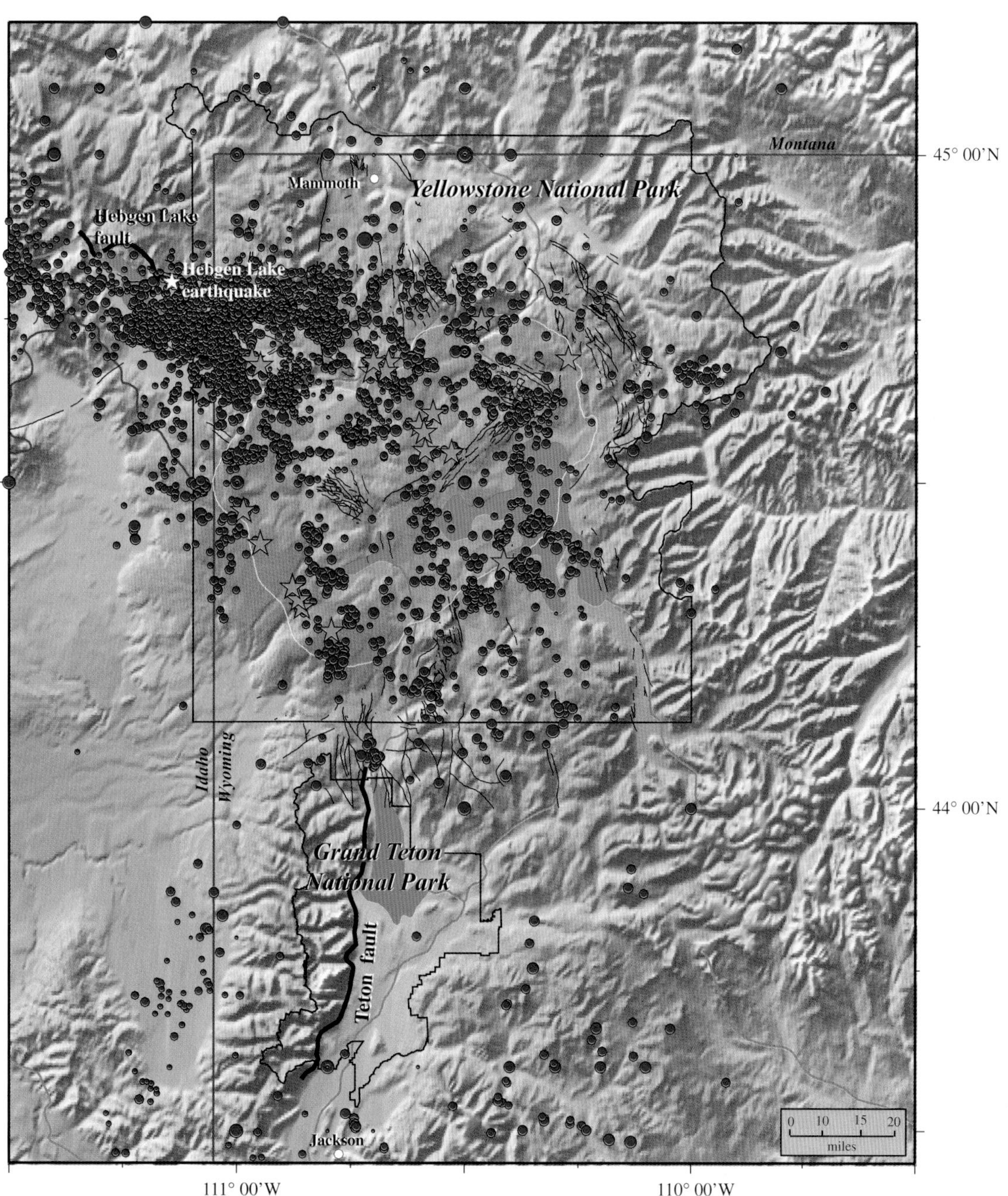

4.9 ∼ *Earthquakes of the Yellowstone–Teton region. Epicenters of earthquakes from 1973 to 1996 are indicated by red dots. Most of the quakes were under magnitude 5. The most intense earthquake activity is in the northwest corner of Yellowstone between Norris Geyser Basin and the Hebgen Lake fault. The Teton fault now is seismically quiet. Active faults are shown as black lines and post-caldera volcanic vents as orange stars.*

transmit the information to the university in Salt Lake City, where the data are used for research and informing the public of the risk of quakes in Yellowstone.

Hundreds to thousands of earthquakes are recorded each year in and around Yellowstone. They range from below magnitude 1 to the magnitude-7.5 Hebgen Lake quake. Most are small—weaker than magnitude 3. Yellowstone's seismicity reveals much about the stresses now active in the region, including the influence of molten rock beneath the surface and the stretching apart of Earth's crust in the Basin and Range Province.

Basin and Range Earthquakes

Even before the hotspot started producing extensive volcanism at Yellowstone about 2 million years ago, big earthquakes rocked the Yellowstone–Teton region for millions of years, helping to shape its spectacular topography. The stretching apart of the Basin and Range Province produced such quakes and continues to do so today.

The Basin and Range Province extends from the abrupt east front of the Sierra Nevada in California to the steep western base of the Wasatch Range in Utah. It includes Nevada, western Utah, and portions of southeastern and eastern California, southeastern Oregon, southern and central Idaho, western Wyoming, and southwest Montana. Driven by movements of Earth's tectonic plates, the crust is stretching apart in an east–west direction in most of this area, and in a northeast–southwest direction in some areas in and around Yellowstone. The different directions of stretching probably relate to differences in how the North American plate of Earth's crust is colliding with crustal plates beneath the Pacific Ocean.

Stretching of Earth's crust creates "normal" faults, the type of fault where ground on one side of the fault moves down and away from ground on the other side. Mountains rise and valleys drop along these faults. Such movement created the striking topography of the Basin and Range: Scores of island-like mountain ranges, each running from north to south, are separated by a series of desert valleys, also elongated in a north–south alignment.

When epicenters from Yellowstone's numerous quakes are plotted on a map, many fall along lines that are oriented from north to south or northwest to southeast. This means the faults that produced the quakes also run along north–south and northwest–southeast alignments. The faults are aligned that way because Yellowstone still is being pulled apart by the stretching that has shaped the rest of the Basin and Range. To accommodate such stretching, faults are formed roughly perpendicular to the direction of the stretching. So are mountains that rise along those faults. In the Yellowstone region, mountains shaped by quakes on normal faults include the steep-sided Teton Range in Wyoming and Idaho and the Gallatin and Madison ranges in Montana.

Unlike most of the Basin and Range Province, Yellowstone's landscape has been significantly altered by volcanic activity in the last 2 million years, hiding many faults. Within the Yellowstone caldera, many earthquake epicenters fall roughly along two northwest–southeast lines. As mentioned in Chapter 3, these lines may represent buried faults along which tall mountains rose until they were destroyed by Yellowstone's first caldera eruption 2 million years ago. The lines nearly coincide with vents through which lava flowed after the big caldera explosions. This suggests Basin and Range stretching of Earth's crust not only triggered mountain-building in the Yellowstone–Teton region, but also influenced the pattern of volcanic activity at Yellowstone.

Earthquake epicenters also fall in a northwest–southeast line in the southern Yellowstone Plateau. This line may represent an extension of the Teton fault buried beneath the lava flow that forms the Pitchstone Plateau. It is not known why this extension of the north-trending Teton fault would bend to the northwest.

Not all of Yellowstone's quakes fit the pattern of falling along north–south or northwest–southeast faults and volcanic vents created by Basin and Range stretching of Earth's crust. North of Madison Canyon, epicenters fall along an unusual east–west line—apparently a fault zone that parallels what may be a deep volcanic vent. There is no good explanation for this east–west alignment.

Volcanic Earthquakes

While many quakes inside the Yellowstone caldera fall along the two northwest–southeast sets of faults and volcanic vents, many others do not. Many of those tremors occur in swarms, which are clusters of quakes that occur within a certain period of time and within a limited area. Quakes in a swarm are similar in magnitude—usually small to moderate quakes without a notable main shock significantly stronger than the rest of the tremors.

Earthquakes within the caldera generally are weaker than quakes outside it, and have not exceeded magnitude 5. Quakes inside the caldera also tend to be centered at shallow depths of only a few miles, while quakes outside it tend to be centered at depths of 10 to 12 miles. Unlike the quakes on the northwest–southeast lines, quake swarms usually are scattered throughout the caldera.

Why are there swarms of small, shallow, scattered quakes inside the caldera? The answer is hot rock underground. Its presence means the cooler uppermost part of Earth's crust is relatively thin inside the caldera. Quakes only occur in relatively cool rock because hotter rock is flexible or ductile and can stretch without breaking. So quakes in the caldera have to occur at relatively shallow depths where rock is cool. A large fault cannot form in such thin crustal rock, so the quakes are small. The epicenters are scattered because Earth's crust within the caldera is highly fractured—not surprising, considering it was blown apart during big caldera eruptions. Quakes tend to happen in swarms in places such as the Yellowstone caldera where underground stress is released on many small faults in highly fractured rock. The movement of hot water and molten rock beneath the caldera only adds to the quake-generating stress caused by Basin and Range stretching of the region.

Geysers Need Earthquakes

If fluid movements can trigger seismicity, the quakes in turn change the behavior of Yellowstone's geysers and hot springs and ultimately may ensure their survival.

A magnitude-6.1 quake rocked Norris Geyser Basin in March 1975, dramatically increasing the flows of some geysers and hot springs. Others decreased or became discolored with gas bubbles. The 1959 Hebgen Lake quake activated many dormant geysers. Yet the 1959 quake and the 1983 magnitude-7.3 Borah Peak quake near Challis, Idaho, 160 miles from Yellowstone, decreased how often Old Faithful erupted. The interval increased to 80 minutes after small quakes in 1998. Geyser activity has been changed by tremors as small as magnitude 4 and by swarms of smaller quakes. A 1978 swarm between Yellowstone Lake and Canyon Junction coincided with an increase in flow from Mud Volcano, raising ground temperatures and killing many nearby trees. Geysers temporarily erupted in the Yellowstone River and in a small lake near Mud Volcano. A magnitude-2.2 temblor in 1998 preceded the revival of Cascade Geyser, dormant since the 1800s, and Vault Geyser, which had not erupted since 1988. Yet Old Faithful erupted a few minutes less frequently after the quake. Old Faithful has been less than faithful over the decades, with intervals between eruptions changing in apparent response to various quakes.

Even quakes far from Yellowstone influence the park's hydrothermal system in unexpected ways. A swarm of tiny quakes in Yellowstone in 1992 was triggered by the magnitude-7.3 quake at Landers, California, some 800 miles away. One theory is that seismic waves from the distant quake briefly compressed rocks at Yellowstone, increasing the pressure of hot water flowing through cracks in the rocks. This increased

4.10 ~ *Elk roam Yellowstone's beautiful Mammoth Hot Springs. These terraces are composed of travertine, which is derived from limestone. (Henry Holdsworth.)*

fluid pressure reduced friction enough to let rocks slide past each other on those cracks, producing small tremors.

Scientists speculate bigger, nearby jolts are required to keep Yellowstone's plumbing system open. The hot-water conduits feeding geysers and hot springs become clogged by minerals that precipitate out of the water, just as waterlines in homes with hard water can become clogged. Frequent quakes break up mineral deposits and create new fractures in rock, allowing hot water to keep flowing to geysers and hot springs. Even Old Faithful could sputter out and die if nearby quakes ever stopped. Despite the frequency of quakes in Yellowstone as a whole, the park contains several geyser basins that went extinct due to the lack of adequate shaking.

The Shifting Shore

As Idaho and western Wyoming drifted over the Yellowstone hotspot during millions of years, the landscape directly above the hotspot bulged one-third of a mile upward. Then, as the crust continued to drift, the high terrain sank again in the hotspot's wake.

For more than the past 2 million years, the Yellowstone Plateau—already high in the Rockies—has been lifted higher on top of the hotspot.

At an elevation of about 8,000 feet, the plateau includes the central portions of the national park, including Yellowstone Lake, Hayden Valley, and the Old Faithful area. Surrounded by peaks of more than 12,000 feet, the plateau is relatively flat, with low rolling hills and ridges covered by dense lodgepole pine forests and open valleys of grass and sagebrush. The flattened topography exists because explosive volcanism destroyed ancestral mountains; the landscape was then smoothed by smaller lava flows, glaciers, and lake sediments.

The heart of this high plateau is the giant volcano known as the Yellowstone caldera. The west half of Yellowstone Lake occupies the eastern part of the caldera. Near the lake's shores, flat areas are perched at elevations ranging from a few yards to a few tens of yards above the present water level. These flat areas, called terraces, are like broad stairsteps. Each terrace is an ancient beach, formed when lake waters were higher than they are now. Since glaciers retreated from Yellowstone about 14,000 years ago, the lake's ups and downs have sculpted terraces at a minimum of eleven distinct elevations. Signs of five of the terraces can be seen all around the lake.

The terraces reveal far more than a record of changing lake levels. They unveil what in geological time is the most recent history of the Yellowstone caldera: how it repeatedly heaved upward and sank downward. The caldera floor rose when fresh molten rock was injected beneath the caldera, and from the resulting buildup and expansion of hot water at shallower depths. The caldera dropped down as magma or hot water either drained away or cooled.

Researchers believe each terrace represents one episode, lasting roughly 1,000 years, during which the caldera floor rose upward, leaving an old beach high and dry, then sank again. The lake terraces reveal that despite ups and downs, the caldera floor generally sank from the time the glaciers receded until about 4,500 years ago. Since then, and despite recent ups and downs, the main part of the caldera has moved upward while its edges sank. Based on estimates of how much magma has been injected beneath the caldera in recent millennia, the caldera's movements suggest hot rock made the middle of the caldera bulge upward while stretching of the Basin and Range Province pulled the caldera apart, making its edges sink.

The Yellowstone Caldera Huffs and Puffs

Like calderas in Italy and California, Yellowstone repeatedly huffs a few feet upward and then puffs downward on a scale of years or decades—movement unnoticed by millions of visitors. That makes Yellowstone what some scientists have called "a giant caldera at unrest."

In the 1970s, scientists noticed trees at the south end of Yellowstone Lake were inundated as the part of the caldera under the north end of the lake rose upward and the part under the south end sank. Precise surveying of benchmarks on the ground showed the Yellowstone caldera rose almost 30 inches from 1923 to the mid-1970s. The ground rose another 10 inches from the mid-1970s to 1984. In other words, the central portion of Yellowstone National Park bulged upward 3.3 feet from 1923 to 1984. Later surveys, including precise measurements using Global Positioning System (GPS) satellites, showed the uplift halted and then reversed, with the central Yellowstone caldera dropping by almost 8 inches from 1985 to 1995. Then, beginning in late 1995 and 1996, images and surveying measurements made by satellites revealed the caldera floor was rising again.

All the measurements taken from 1923 through 1997 have captured—for the first time—a full cycle of how Earth's crust in the Yellowstone caldera is being deformed by the underlying magma chamber and hotspot. Yellowstone's huffing and puffing typifies calderas. Such activity can continue for tens of thousands of years without an eruption. California's smaller Long Valley caldera—the ancient volcano in which the ski town of Mammoth Lakes is located—bulged upward about 15 inches from 1975 to 1983, accompanied by four magnitude-6 quakes and many lesser jolts in the early 1980s. The activity prompted the U.S. Geological Survey to issue a 1982 "notice of potential volcanic hazard," which later was rescinded as the quakes abated. In Italy, the Campi Flegrei caldera near Naples has huffed upward and puffed downward by as much as a dozen yards over several centuries—it rose 6 feet during 1982 to 1984 alone—all without erupting. Numerous earthquakes related to the unrest shook up the 100,000 residents of Pozzuoli, a town in the middle of the caldera.

The Yellowstone caldera's uplift from 1923 to 1984 most likely was caused by molten rock moving upward several miles beneath the caldera, the pressurization of underground hydrothermal reservoirs sealed by minerals and/or the cooling of molten rock. Such cooling could have released gases and hot water that migrated upward and increased the pressure in the shallower hot-water system enough to lift the entire caldera floor. The sinking of the caldera floor starting in 1985 may have resulted from a reduced rate of intrusion of new magma and hot water beneath the caldera, or from molten rock, hot water, and gas draining underground away from the caldera. The renewed uplift of the caldera starting in 1995–1996 could be attributed to the same causes as the 1923–1984 uplift. The Yellowstone caldera is likely to keep huffing and puffing repeatedly for a long time before it bursts into a violent volcanic eruption.

Mystery of the Hebgen Lake Fault Zone

The most prolific of Yellowstone's recorded earthquake swarms did not occur inside the caldera, but just outside its northwestern boundary in late 1985 and early 1986. At its peak in November 1985, the swarm produced hundreds of tremors each day, some with magnitudes up to 4.5. About thirty quakes were strong enough to be felt, prompting some nervous residents to leave West Yellowstone, Montana. The epicenters of the swarm fell on a line extending from the caldera roughly 20 miles northwest toward the Hebgen Lake fault zone, which ruptured during the deadly quake in 1959. The same line also connected with one of the two northwest–southeast lines of epicenters inside the caldera.

It was not known until later, but the 1985–1986 quake swarm started at about the same time the caldera began to sink. Coincidence? Consider some history.

The Hebgen Lake quake was unusual as well as disastrous. Two segments of the fault broke during the quake. Vertical movement on those parallel faults was as much as 22 feet. (See Figure 1.2.) That is extraordinary movement considering that the total length of the faults that broke was less than 25 miles. Put another way, the quake relieved an unusually large amount of stress given the length of the fault rupture. Usually it would take a longer fault to create so much vertical movement. This suggests the Hebgen Lake quake was caused not only by the Basin and Range stretching of Earth's crust but also by the uplift of the entire Yellowstone region due to the hotspot.

The southwest–northeast stretching explains the northwest–southeast alignment of the Hebgen Lake fault zone, the line of epicenters from the 1985–1986 quake swarm, and one of the two lines of epicenters and lava vents inside the Yellowstone caldera. This long crack in the Earth is being pulled open perpendicular to the direction of stretching.

At the same time, the hotspot sends magma upward beneath the region. Some of that magma and overlying hot water intrudes into overlying, vertical faults or cracks, forming vertical walls of molten rock, known as dikes. The two northwest–southeast lines of epicenters inside the caldera likely represent quakes triggered by magma squeezing upward to form two dikes.

The 1985–1986 swarm of quakes could have been caused by magma or hot water draining out of the caldera to the northwest along a long line of dikes and underground faults created by Basin and Range stretching of the region. The close timing of the quake swarm and the sinking of the caldera floor make that a tempting explanation.

The Broken Earth

Why the Tetons Are Grand

5

On a summer morning when the breeze blows cool, it is easy to realize the lakes and sagebrush-covered glacial plains of Wyoming's Jackson Hole sit at nearly 7,000 feet elevation. Yet the altitude of this gorgeous valley is diminished by the view to the west: The precipitous east front of the Teton Range towers above the valley floor, with 13,770-foot Grand Teton and other rugged, snowclad peaks catching the first golden rays of daybreak.

This is one of the most spectacular mountain vistas in America (Figure 5.1). Whether at chill dawn, in glistening light after a torrential afternoon thunderstorm, or during summer evenings when the sun descends behind the jagged Tetons, it is a view that brings solace and peace.

Yet the serene splendor of Grand Teton National Park belies a hidden fury. It is not volcanism, which is concealed beneath the gentle pine-covered Yellowstone Plateau to the north. Instead, this defiant topography was born of seismic disaster as the Teton fault repeatedly and violently broke the earth, producing a few thousand magnitude-7 to -7.5 earthquakes during the past 13 million years.

During each major jolt, Jackson Hole dropped downward and the Teton Range rose upward, increasing the vertical distance between the valley and the mountains by

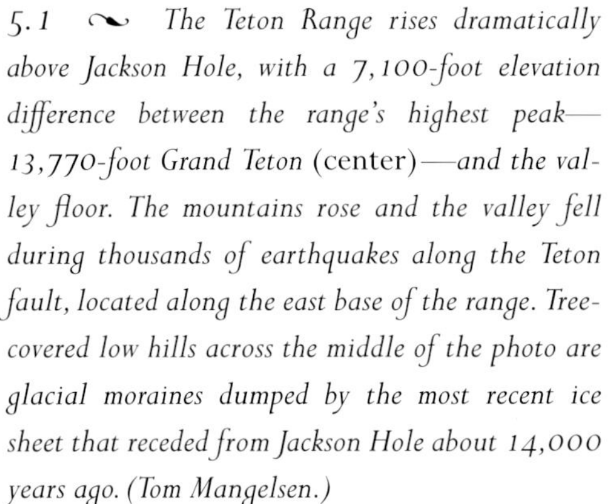

5.1 ~ *The Teton Range rises dramatically above Jackson Hole, with a 7,100-foot elevation difference between the range's highest peak—13,770-foot Grand Teton* (center)*—and the valley floor. The mountains rose and the valley fell during thousands of earthquakes along the Teton fault, located along the east base of the range. Tree-covered low hills across the middle of the photo are glacial moraines dumped by the most recent ice sheet that receded from Jackson Hole about 14,000 years ago. (Tom Mangelsen.)*

3 to 6 feet and sometimes more (Figure 5.2). Now, after 13 million years of earthquakes, the tallest peaks tower almost 7,000 feet above the valley floor.

Actual movement on the fault has been even greater. Jackson Hole dropped downward perhaps 16,000 feet during all those earthquakes. Rock eroded from the Teton Range and other mountains by streams and glaciers filled Jackson Hole with thousands of feet of sediment, disguising how much the valley sank.

Combine the uplift of the mountains and the sinking of Jackson Hole, and the best estimate—although still plagued by uncertainty—is that movement on the Teton fault has totaled 23,000 feet during the past 13 million years.

That is a tiny fraction of Earth's 4.6-billion-year history. Consider the effects of repeated episodes of mountain-building during eons before the Teton fault was born: The oldest rocks high in the Teton Range are 2.8-billion-year-old gneisses and schists and 2.4-billion-year-old granites. They have been lifted as much as 33,000 feet from their subterranean birthplaces—more than the height of Mount Everest (Figure 5.3).

As impressive as that sounds, the power of the Teton fault is revealed by how quickly, in geologic time, it has lifted the mountains. Of the 33,000 feet or so of mountain-building uplift in 2.8 billion years, more than two-thirds of the movement happened only since the modern Teton fault became active 13 million years ago.

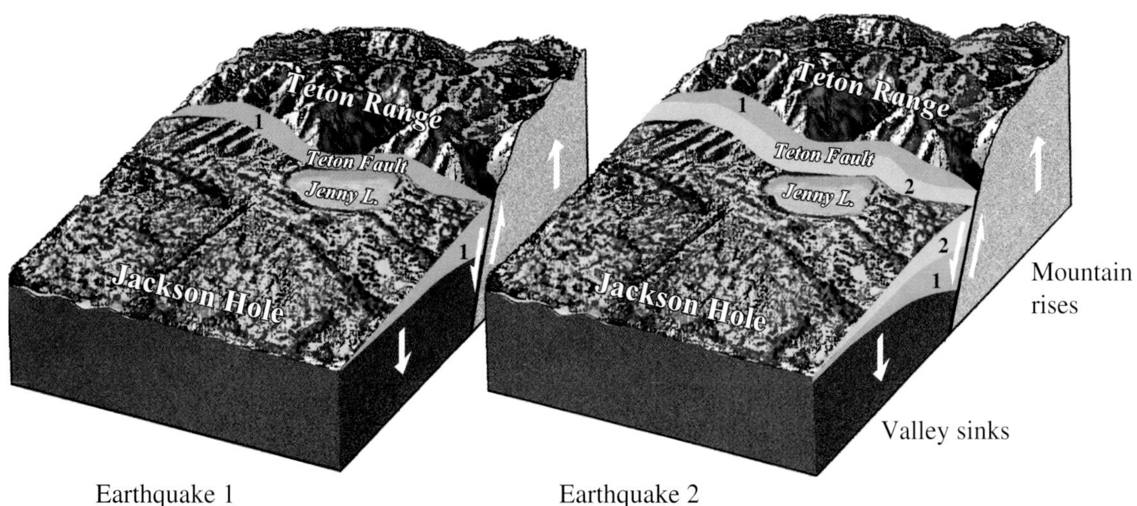

5.2 ~ *Each major earthquake on the Teton fault produced 3 to 6 feet of vertical ground movement, creating a steep scarp or face on the mountain front. Each quake makes the scarp taller, shown here with quakes 1 and 2. The valley floor drops and tilts along the fault as the mountains rise. Sediments eroded from the mountains fill the sinking valley to keep it relatively flat.*

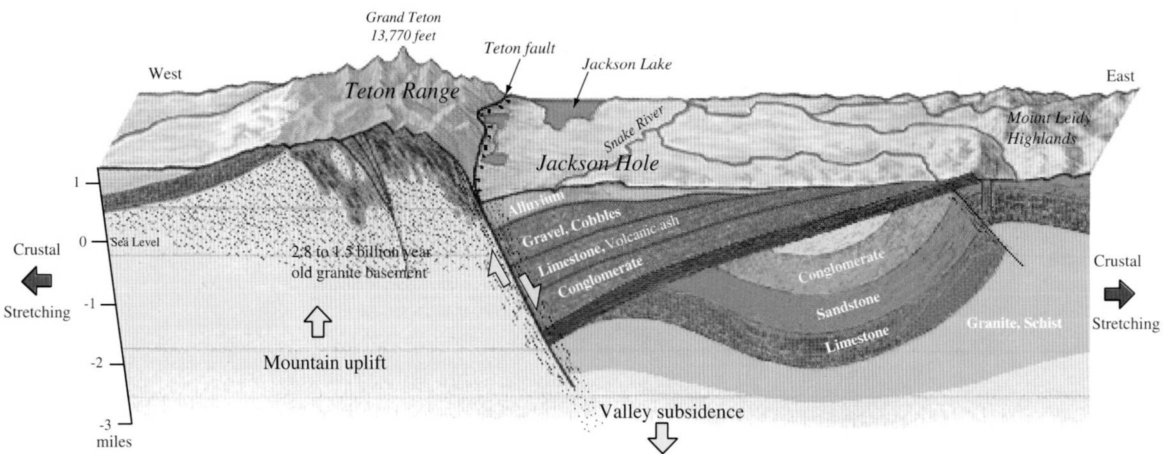

5.3 ~ *Cross section showing the fault-caused westward tilt of rocks on the west side of the Teton Range and beneath Jackson Hole. Total movement on the Teton fault and from earlier mountain-building is about 6 miles, which is the elevation difference between rock layers high in the mountains and their projected location beneath Jackson Hole.*

The seismic power of the Teton fault is not unique. Grand Teton National Park sits within the Intermountain Seismic Belt. The belt is a zone of weakness within the North American plate, one of the great, drifting slabs that make up Earth's crust and upper mantle.

The Rocky Mountains and Colorado Plateau sit east of the seismic belt. To the west is the Basin and Range Province, a region of the interior western United States that is being stretched apart. This stretching is responsible for creating the Teton, Wasatch, Hebgen Lake, and other powerful faults within the Intermountain Seismic Belt.

The belt is apparent on seismicity maps as a diffuse band defined by the epicenters of earthquakes, most of them small to moderate but also some of the strongest in U.S. history: the magnitude-7.5 Hebgen Lake disaster northwest of Yellowstone in 1959 and the 7.3 Borah Peak, Idaho, quake in 1983. The Hebgen Lake quake, centered only 55 miles northwest of the Teton Range, was the largest historic earthquake in the Rockies and the Intermountain Seismic Belt.

The Teton fault is a "normal" fault. Because the Earth is being stretched apart in the region, ground on one side of the fault moves down and away from ground on the other side during earthquakes. It is a different kind of fault than California's San Andreas, which is a "strike-slip" fault where ground on one side of the fault moves horizontally past ground on the other side. Normal faults like the Teton generated the Hebgen Lake and Borah Peak quakes. Violent ground shaking and rupturing of the landscape during those disasters indicate what lies in store for Grand Teton National

Park when the Teton fault eventually lets loose—and for Salt Lake City when the Wasatch fault breaks.

The Teton fault first was described in the 1930s, but detailed studies did not begin until the 1970s, when there was wide recognition that it was capable of generating large earthquakes that could trigger landslides and avalanches, and threaten structures in the region, most significantly Jackson Lake Dam. If a major quake collapsed the dam, the upper 40 feet of water in this large lake would rush down the Snake River, flooding downstream inhabited areas such as Moose—the site of Grand Teton National Park headquarters—and low-lying areas near Teton Village and the town of Wilson. Because of the threat, the dam was strengthened in the late 1980s.

There are large uncertainties in estimating how often big destructive quakes rip the Teton fault. However, an average of roughly once every couple thousand years is a good estimate. The last magnitude-7-plus quake happened sometime between 4,840 and 7,090 years ago. That suggests another seismic disaster in the Tetons—the first to be experienced by modern humans—is overdue.

Yet the 2,000-year "repeat time" for major quakes is only an estimate. And there is debate about the once-popular idea that "characteristic earthquakes" happen like clockwork on certain faults. Different kinds of faults—including the Teton and San Andreas—display evidence that clusters of several strong quakes can occur in a relatively short geologic time, followed by long, quiet periods with no earthquakes. So it is possible the Teton fault is now in such a slumber period, and the next major shake may be thousands of years in the future.

Whatever the uncertainty about the timing of the next big quake, there is little doubt it will happen. The Teton fault is "locked" and represents a "seismic gap"—a stretch of the Intermountain Seismic Belt deficient in quake activity in recent millennia. The quiet period must end, sooner or later, with a major quake that fills in the gap.

"Trois Tetons"

The Tetons were named by lonely French-Canadian beaver trappers in the late eighteenth century. They referred to Grand, Middle, and South Teton peaks as "Trois Tetons" or "three breasts." Nathaniel P. Langford—who explored Yellowstone in 1870 with Henry Washburn and military escort Lt. Gustavus Doane—said of the trappers: "He indeed must have been of a most susceptible nature and, I would fain believe, long a dweller amid these solitudes, who could trace in these cold and barren peaks any resemblance to the gentle bosom of a woman."

John Colter ventured into Jackson Hole shortly after leaving the Lewis and Clark expedition. He entered the valley on its east side and continued north into Yellowstone in late 1807—the first white man known to have visited the region. Many trappers visited Jackson Hole until beavers were trapped out and the fur trade collapsed in the late 1830s. It was left in solitude until the late 1800s, when geologists detoured from their exploration of Yellowstone and ostensibly made the first ascents of Grand Teton. Soon, pioneers began homesteading Jackson Hole—named for a fur trader, Davy Jackson—and established the town of Jackson, Wyoming.

Unlike Yellowstone, which was designated a national park a year after Ferdinand Hayden's 1871 expedition, it took decades of controversy to fully protect Grand Teton National Park, the "Switzerland of America." The Teton Range became part of a national forest in 1908 and much was designated Grand Teton National Park in 1929. Jackson Hole was excluded due to opposition by ranchers who feared for their livelihood and wanted freedom from government regulation. After John D. Rockefeller quietly bought several ranches and donated them to the government, the protected area was enlarged to include Jackson Hole and the north end of the Tetons in 1943. The park and monument were combined into Grand Teton National Park in 1950. The park encompasses the Teton fault, the Teton Range that rose along the fault and the valley of Jackson Hole, which dropped along the fault. The park owes its scenery and existence to an active fault.

Topography of the Tetons

Grand Teton National Park's landscape has been shaped by its active geology (Figure 5.4). The landscape reveals those processes: faulting, other earlier episode of mountain-building, and the relatively recent passage of huge glaciers.

The Teton Range extends 40 miles, from north to south, and is 15 miles wide east to west. Its spectacular eastern front is steep, gaining it world renown among mountaineers. From the loftiest peaks, the range drops eastward to Jackson Hole in only a few miles (Figure 5.5). In contrast, the western side of the Teton Range is a long, gentle slope, taking more than 10 miles to descend from the highest peaks west into Idaho. This striking lack of symmetry betrays the evolution of the Teton Range. It is evidence for the long-lived and still active Teton fault, which runs more than 40 miles from north to south in Jackson Hole along the eastern base of the range. The east front of the Teton Range runs along a fairly straight north-to-south line because the fault cleanly separates the mountains from Jackson Hole.

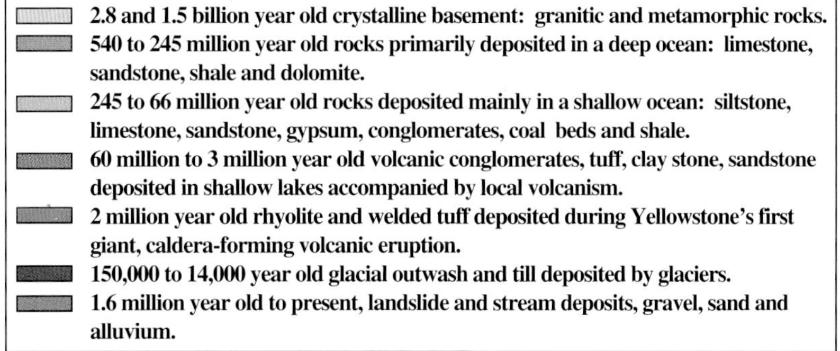

5.4 ~ *Topography and geology of the Teton region. The Teton Range's core is made of 2.4- and 2.8-billion-year-old rocks called gneiss and schist. The youngest rocks are valley floor sediments deposited by glaciers and rivers. The Teton fault (hachured line) extends about 44 miles north–south and separates the uplifted mountain block from the valley floor of Jackson Hole. Volcanic rocks from Yellowstone's eruptions cover the northern part of the range. They were uplifted several thousand feet by the Teton fault.*

The valley of Jackson Hole is bounded by the precipitous Teton Range on the west and highlands on the east. The valley is up to 10 miles wide and 50 miles long, stretching from the north end of Jackson Lake southward beyond the city of Jackson, Wyoming. Jackson Hole is vegetated largely by sagebrush and grass flats, with some areas of cottonwood and aspen, and hills of glacial debris covered by pines and firs.

Imagine the steep east face of the Teton Range extending downward and eastward beneath Jackson Hole, like a big ramp slanted at an angle of between 45 and 60 degrees. This is the Teton fault. The fault intersects the ground surface at the east base of the Teton Range, then dips down and eastward beneath Jackson Hole, eventually reaching a depth of about 10 miles under the valley's east side.

During big earthquakes—jolts measuring 7 or more in magnitude—the valley dropped down and eastward along the ramp-like fault, tilting a bit westward in the process. Meanwhile, the mountains rose up and tilted westward. Over 13 million years, this activity created the abrupt east front of the Tetons and the gentle slope of the range's west side. The Teton Range is a westward-tilted block of rock bounded by the fault on its east side.

Each big earthquake on the Teton fault lifted the Teton Range an average of 3 to 6 feet upward relative to the land east of the fault, creating a small cliff—called a

5.5 ~ *Aerial view of the Teton Range and Jackson Hole looking north. This picture roughly corresponds to the cross section in Figure 5.3. Jackson Hole* (right) *is bounded by the uplifted Teton Range* (left), *which tilts west. Sediment layers under Jackson Hole also dip west against the Teton fault.*

scarp—where the fault intersected and broke the ground surface. Over time, repeated quakes made the cliffs taller. Because glaciers resculpted the landscape until 14,000 years ago, only scarps younger than that are visible now. A prime example is the scarp above String Lake (Figure 5.6).

Of the total movement on the fault from the last 13 million years of earthquakes, Jackson Hole dropped roughly twice as far as the Teton Range rose. Erosion ate away at the mountains. Streams carried eroded rocks, gravel, and sand into the lakes and streams of Jackson Hole, filling the valley and making it relatively flat. This pattern has been observed on other normal faults, including those that produced the 1959 Hebgen Lake and 1983 Borah Peak quakes.

It is easy to visualize why the stretching of the Earth in the Teton region makes Jackson Hole sink lower along the fault during big quakes. It is harder to understand why the mountains rise. An analogy can help.

Imagine cutting a piece of lumber—a 2-inch-by-4-inch stud—in half. But instead of cutting straight down, cut downward at a 45-degree angle. This slope represents the Teton fault. Now put both pieces of wood in a tub of water and watch how they float. The piece of wood that is longer on top represents Jackson Hole. It floats lower in the water because it has more mass on top. The piece of wood that is longer on the

5.6 ~ *The steep slope in shadow, stretching left to right across the middle of the photograph, is the scarp or embankment formed by several magnitude-7 and larger earthquakes on the Teton fault since the glaciers receded 14,000 years ago. The quakes caused 125 feet of vertical movement, creating the scarp in glacial debris above String Lake. (National Park Service.)*

bottom represents the Teton Range. It floats higher because there is more mass on the bottom, and that mass of wood is less dense than the water. In a similar manner, the Teton Range rises upward because the rocks that make up the roots of the mountain range are less dense than underlying rocks.

The lofty crest of the Teton Range is a cathedral of sharp pinnacles, including the highest peak, Grand Teton, at 13,770 feet. To its north is Mount Owen at 12,928 feet and Mount Moran at 12,605 feet. South of Grand Teton is 12,804-foot Middle Teton and 12,514-foot South Teton. This throne of peaks is cut by deep alpine canyons that drop eastward into Jackson Hole. They were carved by glaciation and stream erosion (Figure 5.7).

5.7 ⁓ *The central part of the Teton Range towers 7,100 feet above Jackson Hole due to movement on the Teton fault, which runs along the east base of the range. From left to right are South Teton, Middle Teton, Grand Teton, Mount Owen, and Teewinot Mountain. Grand Teton is topped by 2.4-billion-year-old granites, with 2.8-billion-year-old gneiss on its lower eastern face. The eroded canyons were scoured and the high peaks sharpened by glaciers.*

Year-round snow fills crevices and gullies at the top of the range, which has a dozen glaciers. These glaciers and their larger prehistoric predecessors have scoured and scraped the Tetons, sculpting the fine details of its ragged spires, horns, and precipices.

Faulted Earth, Flowing Waters

The Teton fault also shaped the topography of streams and lake basins, and thus influenced the flow of water. In most mountain ranges, the crest is the highest ridge and also delineates the drainage divide, with water flowing in opposite directions from the crest. Yet the high crest of the Teton Range is 2 to 3 miles east of the range's drainage divide and only a few miles west of Jackson Hole. Canyons rising from Jackson Hole cut through the crest and extend far west of it, so water falling west of the crest actually can flow downhill to the east. This odd effect is a direct result of the rapid uplift of the east face of the Tetons along the Teton fault. Streams on the precipitous east slope have greater power to erode than streams flowing more gently down the west slope. So the east-flowing streams cut canyons that are deeper and extend farther upstream to the west, encroaching on the west slope and diverting its drainage eastward.

Like the west slope of the Teton Range and the rock layers within it, Jackson Hole and the underlying rock layers also tilt gently westward. Most of the valley floor tilts less than 1 degree, which is not easily visible to the naked eye but can be measured by surveying instruments. Part of this tilt is due to debris that washed out of glaciers and is thicker in the middle of the valley than near the west side and the mountains. However, Jackson Hole's west side is lower than its east side primarily because the west side is closest to the Teton Range and Teton fault. During big quakes, vertical movement was greatest near the fault, so the west side of Jackson Hole dropped more than the east side, tilting the valley floor and underlying sediment westward.

The effect of the tilting was to help create a low trough along the west side of Jackson Hole. Water still flows into the trough, helping to explain why a string of scenic lakes sits on the west side of Jackson Hole: Jackson, Leigh, Jenny, Bradley, and Taggart among them.

The Snake River originates in Yellowstone and flows south into the north end of 16-mile-long Jackson Lake. The river flows eastward out of the lake's east side, then turns south and southwest, leaving Jackson Hole at the valley's southern end. The ancestral Snake meandered back and forth across this valley, changing courses repeatedly over time and leaving a series of step-like terraces as evidence of ancient channels.

Streams flowing east out of the Teton Range normally would be expected to flow directly eastward into the Snake River. The lakes also would be expected to drain eastward into the Snake. Instead, the streams either flow southward once they enter Jackson Hole and later join the Snake, or they flow into the lakes, which then drain to the south via Cottonwood Creek. Why? For the same reason the lakes are where they are. The west side of Jackson Hole is lower than the east side because of the fault, and the streams and lakes cannot flow uphill to the east. Jackson Lake is an exception. It drains east toward the Snake River—and did so even before the dam raised the lake level—because tall ridges or "moraines" of glacial debris block the south end of the lake.

The lakes also provide evidence of more recent phenomena that shaped this gorgeous landscape. So do depressions, ponds, and small hills in the sagebrush flats at the north end of Jackson Hole. These features were left by the most recent of a series of glaciers that advanced and retreated over Jackson Hole in three episodes starting perhaps as much as 2 million years ago and ending 14,000 years ago. The glaciers originated high on the Yellowstone Plateau and pulsed north and south. As the glaciers moved south into Jackson Hole, they scoured the landscape, scraping out and deepening lakes and ponds. They also dropped large amounts of rocky debris, creating island-like hills that now are covered by conifer trees and rise above Jackson Hole's sage-filled flats.

Birth of a Fault

The modern Teton fault, which was born as long as 13 million years ago, has existed for a mere trifle in Earth's history. The fault and more recent glaciation shaped the modern Teton Range, but Precambrian rocks form the backbone of the range and reveal ancestral mountains were being uplifted there at least 2.8 billion years ago, only to be eroded during ensuing eons.

Another major period of mountain-building formed the Rocky Mountains roughly 80 to 30 million years ago. It happened because one of the drifting tectonic plates that make up Earth's crust and upper mantle pressed against the west coast of North America and angled downward beneath the continent—a process called subduction. Over millions of years, much of the West was squeezed by the pressure of colliding plates. This compression started influencing the Teton region roughly 60 million years ago.

Such horizontal pressure created faults known as thrust or overthrust faults, in which rocks on one side of a fault are pushed up and over generally older rocks on the other side. Thrust faults generally are gently sloping, with an angle of about 30 degrees. They are the opposite of normal faults, like the Teton fault, in which stretching of the landscape makes rocks on one side of a normal fault move down and away from rocks on the other side. Normal faults tend to have steeper angles of 45 to 60 degrees.

As the West was squeezed to uplift the Rockies, there was extensive thrust faulting in a wide band running north to south along Wyoming's border with Idaho and Utah—an area known as the Wyoming Overthrust Belt. The east–west squeezing of this belt of land was so intense that layers of rock not only were thrust up and over other layers, but folded over them as well—like the folds created in a throw rug if you push one edge with your foot. The petroleum industry has drilled extensively to recover oil and gas trapped in the folds.

5.8 ~ *View of flat-topped mountains of the northern Wind River Range, 60 miles southeast of the Teton Range. More than 34 million years ago, the Teton and Gros Ventre ranges may have been connected as a flat, high plateau. Then erosion started carving the modern Teton Range, which at the time may have resembled this part of the Wind River Range. (David Lageson.)*

East-to-west compression and thrust faulting in the region also made a broad area of western Wyoming arch upward. This area now includes the Tetons and other ranges to the south and east, including the Gros Ventre and Wind River mountains. However, more than 34 million years ago, the Tetons and Gros Ventres were not distinct ranges, but rather, a continuous high plateau (Figure 5.8).

After being squeezed for millions of years, much of the interior western United States began stretching apart. The Farallon plate was a huge block of Earth's crust beneath the Pacific Ocean. It had been pressing against and diving beneath North America's Pacific coast for millions of years, but finally disappeared, completely subducted beneath the continent. The relative motions of the Pacific and North American plates changed. The West no longer was in a vise. The lack of compression eventually allowed hotter rock from Earth's mantle—and possibly from the Yellowstone hotspot when it was under Nevada—to slowly rise upward, making overlying rock in the crust gradually spread sideways.

This lateral stretching or extension created what is called the Basin and Range Province, a wide area of the interior West that is being ripped apart in an east–west direction. That splits open the Earth to create normal faults that run north to south. As the stretching proceeds, valleys drop downward and mountains rise upward along these faults. Like the faults, the mountain ranges and valleys also run north to south. Someone traveling from Utah west through Nevada makes repeated climbs over mountain ranges and descents into valleys. Most motorists do not realize they are crossing another normal fault roughly every 15 to 20 miles. Nor do they realize that during the past 17 million years or more, stretching of the Earth's surface along those faults has totaled more than 200 miles, doubling the distance between the sites of Salt Lake City and Reno.

The Tetons are more lush than the desert ranges that are farther west and southwest and typify the Basin and Range. Yet they share basic topography: a block of mountains, elongated north-to-south, that rise along a normal fault while an adjacent valley drops downward.

The pulling apart of Earth's crust did not affect all areas of the interior West at once. The compression of the West stopped and Earth's crust began relaxing and spreading in Nevada and western Utah about 30 million years ago, but the intensive period of Basin and Range stretching—accompanied by development of steep, normal faults—didn't really get started in those areas until 17 million years ago.

It is difficult to determine when such stretching began influencing the Teton region to create the Teton fault and begin uplifting the modern Teton Range. There is some evidence that, about 34 million years ago, relaxation of Earth's crust began to break up the high plateau in the ancient Teton region, with a valley—ancestral Jackson Hole—dropping down between what are now the Teton and Gros Ventre ranges. However, the best evidence—revealed in ancient sedimentary rock—indicates the Teton fault became active more recently.

Before the Teton fault and modern Teton Range developed, sands, clays, and volcanic ash were deposited on the bottom of lakes in the area. These sediments are known as the Colter Formation and were laid down about 15 to 17 million years ago.

Then the modern Teton fault was born. Over millions of years, movement on the fault made land on the Jackson Hole side of the fault drop downward and land on the Teton Range side rise upward. The land dropped most near the fault, tilting the valley floor. So early earthquakes on the Teton fault also tilted the Colter Formation until the rocks were inclined at an angle of 15 degrees.

While this was happening, no other sediments were deposited on top of the Colter rocks. Then, sometime before 10 million years ago, part of the Teton area was inundated by a big lake, an ancestor of Jackson Lake. Sediments deposited on the lake bottom became limestones and claystones mixed with volcanic debris—a rock layer now known as the Teewinot Formation. The Teewinot sediments formed flat layers on top of the tilted Colter layers. This type of boundary between the two rock formations is called an angular unconformity, and geologists know it represents a period of time during which no sediments were deposited.

The Teewinot Formation's age has been dated at 10 million years. The period during which no rocks were deposited probably lasted no more than 3 million years. So the best estimate is that the Teton fault became active, started generating major earthquakes, and began tilting the Colter Formation about 13 million years ago. That is when the most recent and ongoing period of mountain-building began uplifting the modern Teton Range.

Jackson Hole's Modern Earthquakes

There have been no major earthquakes on the Teton fault in historic time. From the early 1900s to the early 1930s, the small, isolated population of Jackson Hole reported numerous mild to moderate earthquakes. Several quakes up to about magnitude 5 were centered just east of Jackson Hole beneath Gros Ventre Canyon, and some caused minor damage. Several residents were thrown from their beds by a quake the night of January 26, 1932. The jolt also cracked some foundations and plaster in area homes.

After about 1933, the number of quakes felt in the Teton region decreased sharply, although a 1948 jolt damaged some log buildings. In 1959, the valley was rocked by the magnitude-7.5 Hebgen Lake quake, which was centered about 60 miles northwest of Jackson Hole. Since then, small to moderate shocks have been reported in southern Jackson Hole and the Gros Ventre Range. (See Figure 5.9.)

Installation of portable seismometers in the 1970s and permanent sensors in the 1980s allowed researchers to measure quakes too small to be felt. Epicenters of many of these small quakes form a pattern that extends from Jackson Hole northeast across the Gros Ventre Range. There also have been small quakes with epicenters stretching from north of Jackson Lake north into the Yellowstone Plateau. These quakes may have been generated by an extension of the Teton fault that was covered by volcanic rocks in the last 2 million years. Other small, shallow quakes have been centered in a north-to-south belt near the south end of the Teton fault—but not on the fault itself—and in two clusters in the south end of Jackson Hole between Jackson and Moose. Several of those jolts were felt in Jackson in the late 1980s.

Throughout modern history, there has been a striking absence of quakes centered beneath the northern two-thirds of Jackson Hole. This quiet area extends at least 30 miles, from Moose to north of Jackson Lake. This quake-free zone coincides with the northern Teton fault. The seismic quiet suggests the Teton fault now is locked but slowly is accumulating underground stress that someday will be unleashed during a disastrous earthquake.

Shaking, Sliding, and Flooding

Major quakes are not the only seismic activity capable of disaster in Jackson Hole. Decades before the 1959 Hebgen Lake quake caused a massive and deadly landslide in Montana's Madison Canyon, much weaker quakes may have triggered an even larger landslide in Gros Ventre Canyon, a few miles east of Jackson Hole. The Gros Ventre slide (Figure 5.9) was a world-class landslide, a third larger in volume than the huge slide in Madison Canyon.

For several years in the early 1920s, persistent small quakes rattled the eastern side of Jackson Hole, according to personal accounts of residents living in Moran and Gros

5.9 ~ *The 1925 Gros Ventre slide (also known as the Lower Gros Ventre slide) sent 50 million cubic yards of rock cascading off the south side of Gros Ventre Canyon. (W. B. Hall and J. M. Hill.)*

Ventre Canyon. The canyon runs east to west into Jackson Hole. The number of small quakes increased significantly in the spring of 1925. The winter and spring had been unusually wet, and the ground was saturated at the time. Billie Bier, a Gros Ventre Canyon cowboy, spotted new springs on the south side of the canyon. He wrote:

> I have noticed that and cannot see where the water can be going. . . . The time will come when the entire mountain will slip down. . . . Tremors that are coming so often are going to hit the right time when the mountain is the wooziest, and down she will come.

The late W. C. "Slim" Lawrence, a longtime resident of Moran, was in the dining room of Teton Lodge in Moran at about 8 P.M. on June 22, 1925. Years later, he recalled how the building suddenly rocked during an earthquake that "was strong enough to scare the cook very much, as he came running out of the kitchen. A few dishes fell on the floor and he would not go back in the kitchen that night." Based on various witness reports, the quake probably measured about 4 in magnitude.

The next morning, farmer Guil Huff was plowing his field above Gros Ventre Canyon when he noticed cracks, seeping water, and house-sized areas where the ground slumped. About 4 P.M., while riding horseback looking for cattle in the canyon, he was surprised to see the sudden collapse of the 30-to-40-foot-high riverbank on the south side of the Gros Ventre River. He later recalled:

> Then with a rush and a roar came the entire side of a mountain, spreading out in a fan shape and rolling forward with great speed. I turned my horse and rode with all possible speed up the river, needing to change my course twice in order to keep away from the on-rushing mass of rock, trees and earth. It reminded me of a flood of water, and only the good horse upon which I was mounted prevented me from being buried. The whole thing was over in about a minute and a half.

The Gros Ventre slide of June 23, 1925, sent 50 million cubic yards of rock rushing down the south side of Gros Ventre Canyon. This compares with the 37-million-cubic-yard Madison Canyon slide triggered by the Hebgen Lake quake. The Gros Ventre slide dropped 2,100 feet vertically while traveling almost 1.5 miles. It was 2,000 feet wide. The speeding rock and debris crossed the river and climbed 350 feet up the north side of the canyon. When the dust settled, a 225-foot-deep pile of slide debris dammed the Gros Ventre River, creating 3-mile-long Lower Slide Lake, which flooded some homes and a ranch.

No one was killed by the slide, but disaster was not over. The double whammy concluded almost two years later during another wet spring. Lower Slide Lake filled up and started overflowing the natural dam. The dam's upper 50 feet abruptly gave way on May 18, 1927, unleashing a flood that swept westward 4 miles down the canyon, inundating the town of Kelly with water as deep as 15 feet. The homes and barns of Kelly's eighty residents were destroyed. Fertile fields were washed away and covered by gravel and other flood debris. Six people drowned.

Fifteen miles downstream from Kelly, a few miles after the flood gushed into the Snake River, up to 6 feet of water flooded the town of Wilson. Hundreds of farm animals died. Farther downstream, water levels rose as much as 50 feet in narrow stretches of the Snake River near Hoback Junction, a dozen miles south of Jackson. The next day, the floodwaters covered lowlands at Idaho Falls, 135 miles downstream from Lower Slide Lake. When it was over, damage from the Gros Ventre flood reached $500,000—millions in today's dollars.

Although the evidence is equivocal, it is reasonable to believe earthquake shaking, vulnerable rock in Gros Ventre Canyon, and ground saturated by wet weather conspired to produce the Gros Ventre slide and subsequent flood.

Volcanism and the Tetons

Although the Teton Range and Jackson Hole have been shaped largely by movement of the Teton fault during the past 13 million years and by glaciation within the past 2 million years, volcanism also played a role.

During the past 50 million years—long before the arrival of the Yellowstone hotspot—volcanoes have been active intermittently in the Teton–Yellowstone region, erupting large volumes of lava, ash, and other volcanic debris. Much of the older volcanism—including the eruptions that produced the Absaroka Range in northeast Yellowstone 50 million years ago—likely stemmed from the crustal plate that was diving or subducting under the western coast of North America. The subducting plate not only squeezed the West to build the Rockies, but melted as it angled under the continent. Molten rock rose to produce eruptions. The same process in modern time produced eruptions of Mount St. Helens, Lassen Peak, and other Cascade Range volcanoes in the Pacific Northwest. However, some scientists have argued subduction cannot explain the ancient volcanism in Wyoming. Instead, they argue that once the Rockies were squeezed upward, the relaxation and spreading apart of Earth's crust triggered partial melting of rock within the crust, leading to volcanic eruptions.

Volcanic rocks from eruptions about 9 million years ago are found near the south end of the Teton Range, particularly in the Gros Ventre Buttes just northwest of the city of Jackson. At the time, the Yellowstone hotspot was well to the west beneath the Snake River Plain. It is possible that some molten rock from the hotspot oozed eastward as it rose upward, producing the 9-million-year-old volcanism in the Teton region.

Volcanic rock known as Conant Creek tuff is found on the north and west sides of the Teton Range and on Signal Mountain in Jackson Hole. The tuff is solidified volcanic ash more than 4 million years old. The ash probably fell from the sky on the Teton area after it erupted from the giant Heise caldera, which was on the Snake River Plain directly west of the Teton Range and northwest of Idaho Falls. This caldera marks the likely location of the Yellowstone hotspot more than 4 million years ago, or some 2 million years before the hotspot arrived beneath Yellowstone.

The most prevalent volcanic rock visible in the Teton region is the Huckleberry Ridge tuff—the thick layer of light gray volcanic ash deposited when the Yellowstone caldera first exploded 2 million years ago. It caps the northern end of the range. The ash swept south from the giant caldera, flowing in dense clouds to the north end of the Teton Range. Then it kept flowing southward at least 40 miles down the west and east sides of the range, burying Jackson Hole. Airborne ash fell on the mountains.

It may seem odd, but in addition to creating a volcanic disaster in Yellowstone and the Tetons, the eruption 2 million years ago provides clues to future seismic disasters. That is because the Huckleberry Ridge tuff is a geological marker for measuring past movements on the Teton fault.

The Timing of Big Earthquakes

Rocks do not lie, but they can tell different stories. The history of earthquakes on the Teton fault and hints about its future behavior are recorded in the Huckleberry Ridge tuff and other rocks of the Teton Range and Jackson Hole. Trying to read that history is tricky and filled with uncertainties.

After the Yellowstone caldera's first cataclysmic eruption 2 million years ago, much of the surrounding landscape—including the Teton Range and Jackson Hole—was blanketed by the volcanic ash and debris that hardened to become the Huckleberry Ridge tuff. Over time, powerful earthquakes continued to break the Teton fault. The thick layer of tuff was broken too. Huckleberry Ridge tuff west of the fault rose with the mountains, while the part of the tuff layer east of the fault sank with Jackson Hole.

Today, the Huckleberry Ridge tuff in the Teton Range sits about 13,000 feet higher than the same rock layer buried beneath Jackson Hole. In other words, during the past 2 million years, earthquakes on the fault made the valley drop and the mountains rise by a total of 13,000 feet. If earthquakes produced 13,000 feet of movement on the fault, and if we make a reasonable assumption that each major earthquake produced an average of 6 feet of movement, this means there were about 2,200 major quakes during the past 2 million years, or one big jolt roughly every 900 years.

That is just one estimate. The vertical separation of much older rock layers broken by the fault suggests there has been a maximum of 23,000 feet of vertical movement on the fault—or roughly 3,800 major quakes—within the past 13 million years. That works out to a magnitude-7 quake every 3,400 years.

Yet another story of the Teton fault's quake history is told by younger rocks. During the 14,000 years since major glaciers receded from Jackson Hole, quakes repeatedly broke apart ridges of glacial debris, called moraines, sitting on top of the Teton fault. As ground on one side the of the fault rose during quakes and ground on the other side dropped, the movement created small cliffs or banks called scarps. By measuring the heights of scarps and dating bits of wood that tumbled off scarps right after big quakes and then were buried by other debris, the sizes and dates of the last two major quakes on the Teton fault have been estimated. The most recent one measured about magnitude 7 and rocked the region sometime between 7,090 and 4,840 years ago. The second most recent quake happened about 7,900 years ago and registered about magnitude 7.3. Those two quakes within the last 7,900 years mean an average of a major jolt roughly once every 4,000 years. However, the evidence raises the possibility that the two last quakes occurred close together—namely, 7,090 and 7,900 years ago.

The height of the scarps also can tell us about quakes since the glaciers receded 14,000 years ago. They indicate there were nine big quakes between 14,000 and 7,900 years ago, or one every 680 years during that period.

We now have arrived at four estimates of how often big quakes ruptured the Teton fault during different time intervals: once every 680, 900, 3,400, or 4,000 years. What are we to make of these seemingly contradictory stories told by rocks along the Teton fault?

First, the different estimates are based on evidence found along different parts of the fault. So they indicate major quakes happen at different rates on different parts of the fault. The data reflect frequent prehistoric quakes near the north end of the range within the past 2 million years. It may not be coincidental that the Yellowstone hotspot produced its first cataclysmic caldera eruption 2 million years ago. It is possible, although speculative, that the arrival of the hotspot uplifted the ground not only in

Yellowstone but also farther south, adding stress that made quakes strike more often on the northern end of the Teton fault than on the fault's other segments.

The second lesson from these seemingly conflicting numbers is that another big quake may be long overdue. The rocks tell us that big quakes happen every 680 to 4,000 years. A nice but rough average would be approximately one big quake every 2,000 years. No matter what estimate you prefer, it has been at least 4,840 and possibly 7,090 years since the last disastrous earthquake, suggesting the Teton fault could produce another one at any time.

The data also provide a third, possibly contradictory lesson: Major quakes on the Teton fault clearly do not occur at a constant rate over time. There are periods of more frequent quakes, and periods of less frequent ones. The rocks reveal strong tremors were much more frequent during the past 2 million years than during the preceding 11 million, perhaps reflecting the hotspot's arrival beneath Yellowstone. The scarps tell us that quakes were even more frequent from 14,000 to about 7,900 years ago, and then suddenly became less frequent. Perhaps the massive weight of the glaciers suppressed movements on the fault. When the glaciers melted away 14,000 years ago, the mountains again were able to rise up and the valley to drop down, producing a cluster of nine strong jolts between 14,000 and 7,900 years ago.

No one can say why 4,840 to 7,090 years have passed since the Teton fault last unleashed its seismic fury. The conventional wisdom is that the next big quake is overdue. However, there is a possibility that big quakes routinely happen in clusters, followed by thousands of years of seismic peace. If that is true, no one knows how long the quiet periods last.

Whether the Teton fault is now in a dormant period or overdue for the next strong quake, there is no doubt the Basin and Range Province still is stretching apart, and stress must be building on the fault. Despite huge uncertainties over timing, sooner or later another disastrous quake will raise the Teton Range and lower Jackson Hole a bit more, continuing the violent process that created and maintains the beauty and magnificence of Grand Teton National Park.

Ice over Fire

Glaciers Carve the Landscape

6

Yellowstone, the Tetons, and Jackson Hole were shaped by multiple catastrophes. Huge volcanic eruptions and powerful earthquakes played major roles. Finishing touches were added by another kind of calamity: A rare global Ice Age produced gigantic glaciers that buried the landscape with ice two-thirds of a mile thick in places. The glaciers carved mountains, canyons, and lake basins. They dumped large piles of debris and redirected the flow of rivers. The Yellowstone–Teton region is a world-class example of how land was reshaped by glaciers during what is known as the Pleistocene Ice Age.

The Ice Age was not a single glacial period, but many intermittent cold spells interspersed with warmer periods during which the ice melted. The timing of major glacial periods is notoriously uncertain. Although continental ice sheets did not quite reach as far south as Yellowstone, a regional icecap and large glaciers covered the Yellowstone–Teton country (Figure 6.1) during three major episodes of at least the past 300,000 years—and perhaps the past 2 million years. The last of these big glaciers retreated about 14,000 years ago, although some argue they did not recede until 10,000 to 12,000 years ago. Today, small glaciers in the Teton Range are found only above 10,000 feet.

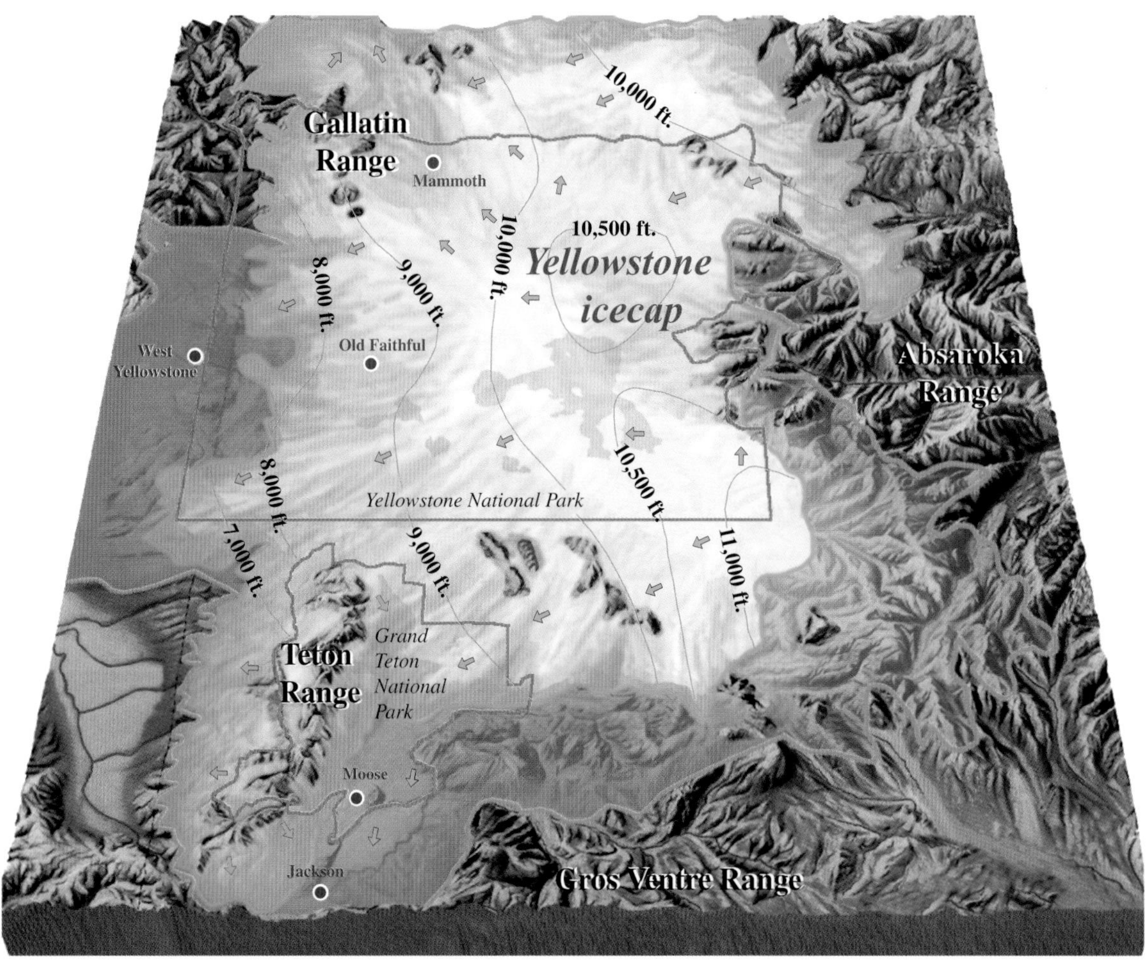

6.1 ~ *A 3,500-foot-thick icecap formed atop the Yellowstone Plateau at various times during three major glacial periods, beginning as early as perhaps 2 million years ago. During the most recent or Pinedale glaciation, roughly 50,000 to 14,000 years ago, ice (light blue) flowed south from Yellowstone and terminated in northern Jackson Hole. During the second oldest glacial stage—the dates of which are debated—ice (darker blue) was up to 2,000 feet thick in northern Jackson Hole and covered the entire valley. Small arrows show the directions of ice flow. (Adapted from Locke, 1995; Pierce, 1979; and Smith, 1993.)*

During each major episode, most of Yellowstone National Park was buried beneath an icecap as much as 3,500 feet thick, among the largest in the ancient Rocky Mountains. Gigantic masses of ice flowed down from the high Yellowstone Plateau, carving and scouring the Earth's surface, diverting and damming rivers into their present forms, steepening mountain fronts, and deepening lakes. The ice helped sculpt the Grand Canyon of the Yellowstone (Figure 6.2). More than anything, the thick ice

6.2 ~ *The Yellowstone River cut down 800 feet to form the Grand Canyon of the Yellowstone. The bright yellow to red rocks of the canyon walls are made of 630,000-year-old solidified lava highly altered by exposure to chemicals, hot water, and steam. Erosion was accentuated by glaciers along with increased water flow as the ice melted. The 310-foot-tall Lower Falls formed where resistant lavas overlie less resistant rock. (Robert B. Smith.)*

6.3 ~ *During the last glacial period, the Burned Ridge Glacier flowed into Jackson Hole from the northeast, followed thousands of years later by the Jackson Lake Glacier, which flowed off the Yellowstone icecap. These and earlier glaciers dumped rock and sediment to form long ridges called moraines. Smaller alpine glaciers descended from the Teton Range.*

scraped Yellowstone's volcanic topography, further smoothing the plateau and helping to excavate the basin occupied by Yellowstone Lake.

Jackson Hole became a rendezvous of glaciers converging from the north, northeast, and west (Figure 6.3). Ice up to 2,000 feet thick scooped out the valley floor. The glaciers left tall ridges of rocky debris now covered by lush conifer forests. Such ridges, called moraines, helped shape Jackson Lake.

Alpine glaciers formed in the Tetons, and the advancing ice gouged deep into bedrock, creating U-shaped canyons where streams once had produced V-shaped canyons. Granite, Death, Avalanche, Cascade, and other canyons took their modern U-shaped forms this way. The glaciers sharpened the spires and ragged peaks of the Teton Range, smoothed out ridges, and polished rock faces to a sheen. Debris from glaciers dammed streams to form high alpine lakes.

The climate warmed and the glaciers retreated some 14,000 years ago. Staggering volumes of melting ice produced flooding, massive debris slides, and erosion, especially when ice dams broke abruptly. Streams and rivers such as the Snake, Madison, and Yellowstone gushed sediments onto broad valley floors. Repeated episodes of flooding cut stairstep-like terraces seen along the Snake River in Jackson Hole and the Madison River near West Yellowstone, Montana.

The advances and retreats of the glaciers put an imprint on the distinctive and spectacular scenery at Yellowstone and Grand Teton national parks.

What Are Glaciers?

Glaciers are large, moving masses of ice that form when snow accumulates winter after winter and is compressed until it forms ice. They exist at elevations where temperatures remain cold enough throughout the year to allow snow to compact into permanent ice. During cooler climate periods, the ice tends to accumulate and glaciers advance, while in warmer periods the ice tends to melt and glaciers retreat. There are exceptions—climate warming can increase precipitation in mountains to make glaciers grow.

The largest ice masses blanket continents such as Antarctica and Greenland, cover entire mountain ranges, or extend miles across large valleys. They are called icecaps or ice fields. Smaller ice masses called alpine glaciers flow down mountain valleys and are contained within those valleys. They take their name from Europe's most glorious mountain range, the Alps.

Glacial ice flows downhill under the influence of gravity, normally several feet per year, but sometimes several feet daily. The ice deforms as it flows. The glacier remains mostly continuous, but the ice cracks as it flows over irregular underlying rocks, creating deep crevasses on the glacier.

A moving glacier is massive enough to break off protruding rocks as it gouges and scours the landscape. Boulders, rocks, gravel, sand, and finely ground rock "flour" become entrained in the ice. Glaciers act like conveyor belts, carrying rock debris within and atop the flowing ice mass.

The boulders and other debris are dropped at the front edge of the glacier to form a "terminal moraine" and at the sides to create "lateral moraines." Moraines are large, hummocky ridges that can reach heights of more than 1,000 feet, although those in Jackson Hole have heights up to a few hundred feet.

Streams often flow atop and beneath glaciers. Fine rock ground into flour by glaciers can turn such streams milky white. When carried into lakes, the rock flour can catch sunlight and give the water a deep milky blue appearance.

The Big Chill

Yellowstone and the Tetons were not alone in the time of glaciers. The Pleistocene epoch—from 1.6 million years ago until roughly 10,000 years ago—was marked by worldwide climate cooling and glaciation, with global temperatures averaging 7 degrees Fahrenheit below today's. Yellowstone's high elevation atop the hotspot may have allowed glaciation to start there even before the Pleistocene, although this is debatable.

Giant ice masses several thousand feet thick advanced over most of northern Europe and northern North America, including Canada, Alaska, and the northern tier of states. The continental ice sheet in North America advanced into northernmost Idaho and Montana, but did not reach Yellowstone. However, large regional icecaps covered Yellowstone, the Tetons, and the rest of the Rocky Mountains, as well as the Sierra Nevada in California.

The geological record reveals two other periods of extended glaciation in Earth's history—the first sometime between 600 and 800 million years ago and the other about 200 to 300 million years ago. The more recent Pleistocene cooling that shaped the Yellowstone–Teton region not only reveals clues about relatively rare events in Earth's history but helps us sort out natural and pollution-caused global climate changes.

Scientists still do not understand what produces continental-scale glaciation. One hypothesis is that the slow movements of Earth's crustal plates may influence climate by changing ocean circulation, sea levels, and precipitation patterns, but Earth's drifting plates did not undergo a major rearrangement during the Pleistocene epoch.

Climate changes also can be explained by the rise of mountains that alter wind patterns. As the Indian plate of Earth's crust smashed into the Eurasian plate to lift the mountains of the Himalayas, monsoon rains increased. Changes in Earth's rotation and orbit also might change climate by altering the heating of the planet's surface. However, none of these mechanisms has been proven as the cause of global cooling during the Pleistocene epoch.

Large volcanic eruptions can contribute to global cooling by darkening the sky for years with airborne gases, aerosols, and ash. The 1883 eruption of Indonesia's Krakatau volcano in the southwest Pacific spewed airborne particles that obscured sunlight and radiated heat back into space, lowering global temperatures by 2 to 4 degrees Fahrenheit. But even huge eruptions such as those at Yellowstone were rather short-lived compared with the 1.6-million-year Pleistocene Ice Age, and their effects, while catastrophic, apparently did not approach the scale needed to plunge the planet into an Ice Age.

A Land of Ice and Fire

If glaciation in the Yellowstone–Teton region did begin as long as 2 million years ago—a debatable contention—it is tempting to note the Yellowstone caldera first exploded 2 million years ago as well. Yet there is no evidence to tie the eruption to such severe climatic change.

Nevertheless, interactions occurred between the glaciers and the Yellowstone hotspot beneath them—giving rise to the notion of the Yellowstone–Teton region as a land of ice and fire.

On a relatively small, local scale, meltwater from ice seeped underground to trigger hydrothermal or phreatic eruptions of steam and hot water, producing craters up to 4,000 feet wide. There are ten known areas in Yellowstone with hydrothermal explosion craters, some in glacial debris. There probably were many more such blasts. Evidence of them has been erased. No one knows if any of the blasts broke through the overlying glacial ice, although it seems likely.

It is hard to imagine thick glacial ice melting completely as it flowed over even a major hydrothermal basin. But heat in a geyser basin may have melted enough overlying ice in some cases to form lakes. If the ice holding back such a lake broke and the lake suddenly drained, the loss of confining pressure might allow an underground hot water reservoir to erupt violently.

On a larger scale, eruptions of rhyolite lavas have happened at Yellowstone caldera after each of its three major caldera explosions and as recently as 70,000 years ago. So it is likely that eruptions occurred when glaciers were present. There are places in Yellowstone where lava apparently was deposited in contact with ice. We do not know the form or scale of such interactions between glaciers and lava flows, but modern eruptions elsewhere give us some good ideas.

Colombia's Nevado del Ruiz volcano erupted steam and ash for two months in 1985 before a major eruption melted ice on the mountain, unleashing a catastrophic mudflow that killed an estimated 23,000 people.

Like Yellowstone, Iceland sits atop a hotspot, albeit an oceanic rather than continental hotspot. In October 1996, Iceland's Loki volcano erupted through a long, underground fissure beneath Vatnajokull Glacier, considered Europe's largest glacier. The eruption initially produced a plume of steam, ash, and gas almost 2 miles tall. Hot rock melted some overlying ice. The water pooled beneath the 2,000-foot-thick glacier. In early November, the glacier ruptured, releasing millions of gallons of water in raging torrents that flooded the countryside, washing away roads and bridges in a remote area of the country's south coast. Another eruption dumped ash on a 3-mile stretch of the glacier.

Mount Rainier—the highest peak in the Cascade Range—is viewed as one of North America's greatest geologic hazards because of the flood threat posed to areas near Seattle and Tacoma if a major eruption melted Rainer's glaciers. River valleys dozens of miles from the volcano were inundated by mud and meltwater after prehistoric eruptions at the 14,410-foot-tall mountain. Those areas now are filled with Seattle suburbs and towns like Auburn, Kent, Puyallup, and Sumner.

In addition to the likelihood of simultaneous eruptions and glaciers at Yellowstone, there was a larger-scale but more subtle interaction between the hotspot and the ice. During the past 2 million years, the hotspot made a 300-mile-wide area centered on Yellowstone bulge upward an extra 1,700 feet, pushing the plateau to 8,000 feet. This extra elevation could have accentuated the effects of global cooling during the Ice Age, chilling the Yellowstone Plateau 8 degrees Fahrenheit compared with surrounding areas and making the Yellowstone icecap larger and thicker than it would have been otherwise.

The presence of an icecap at various times during the Pleistocene likely changed prevailing winds, which today flow from the southwest "up" Idaho's Snake River Plain and onto the higher Yellowstone Plateau. Indeed, the Snake River Plain in some places has a thick layer of soil called loess, which is fine, windblown glacial sediment. The loess indicates that when the icecap was present, rock was ground into fine powder by glaciers, then blown from Yellowstone, Jackson Hole, and nearby mountains southwest onto the Snake River Plain.

Time of the Glaciers

At least three major periods of glaciation buried much of the Yellowstone–Teton region beneath thick ice, with glaciers advancing and retreating repeatedly during each period. There is no consensus on their dates.

1. The first stage, often called the Buffalo stage, may have begun as early as 2 million years ago and ended 1.3 million years ago. Many researchers doubt such early dates for this stage, and argue it began 250,000 or 300,000 years ago and ended 180,000 years ago. Relatively little is known about this early period of glaciation because its influence on the landscape has been erased by more recent glaciers, erosion, and volcanism.

   Hints are found in northern Jackson Hole, where the glaciers dropped chunks of obsidian, a glassy volcanic rock, carried from Yellowstone. The glaciers apparently covered the Beartooth and Absaroka ranges and the Yellowstone Plateau, moved south into Jackson Hole, and flowed over some of the Teton Range. Ice also flowed from the Tetons and Gros Ventre Range into Jackson Hole, where it merged with the south-flowing ice. This major ice sheet continued down the Snake River drainage, perhaps as far as eastern Idaho.

2. The Bull Lake stage variously has been dated as having lasted from 160,000 to 130,000 years ago, from 125,000 to 45,000 years ago, or from 80,000 to 35,000 years ago. Signs of Bull Lake glaciation are found through most of the Yellowstone–Teton region. It is named for glacial deposits at Bull Lake on the east side of the Wind River Range.

   The Bull Lake glacial stage was marked by perhaps ten or more glacial advances and retreats. Signs of this glacial stage can be seen today. The ice deposited rocks and boulders high on mountainsides. It scraped bedrock to create striations on the rock. And it dumped glacial debris that now is buried beneath younger glacial and volcanic rocks. Bull Lake glaciers probably were more extensive than those of the earlier stage. The Yellowstone Plateau, West Yellowstone area, and Absaroka, Teton, and Gros Ventre ranges mostly were covered by ice. From north to south, the ice stretched more than 125 miles, from the Absarokas south through Yellowstone and Jackson Hole to well south of the town of Jackson. Jackson Hole was entombed beneath 2,000 feet of ice. Rocky debris dropped by the ice now caps Signal Mountain, which is just southeast of Jackson Lake and rises several hundred feet above Jackson Hole's valley floor. Such debris also forms the ridge named Timbered Island, southeast of Jenny Lake (Figure 6.4).

3. The Pinedale stage, which began sometime between 50,000 and 25,000 years ago and lasted until the end of the glacial age about 14,000 years ago, and perhaps more recently. Pinedale glaciation is the best known of all stages because its effects are prominent throughout the Yellowstone–Teton region.

6.4 ~ *Timbered Island is a narrow, pine-covered glacial moraine southeast of Jenny Lake. The hill, which extends north–south, was produced by the second of three stages of glaciation in Jackson Hole. As a giant glacier flowed south, it dumped rocky debris along its sides, forming lateral moraines, including Timbered Island.*

An icecap as thick as 3,500 feet covered the Yellowstone Plateau, Absaroka Range, and much of the Teton Range. This icecap extended roughly 60 miles east–west, from the Absarokas to West Yellowstone, and another 80 miles north–south, from Montana south across Yellowstone to the northern Tetons and northern Jackson Hole. The surface of the icecap attained elevations as high as 10,500 feet. It covered the Yellowstone Plateau's highest peaks, including 10,243-foot Mount Washburn. In the Absaroka and northern Teton ranges, only peaks above 10,500 feet poked above the ice. It would have been a remarkable sight. Imagine flying over Yellowstone and seeing an ice field covering an area three-quarters the size of Connecticut.

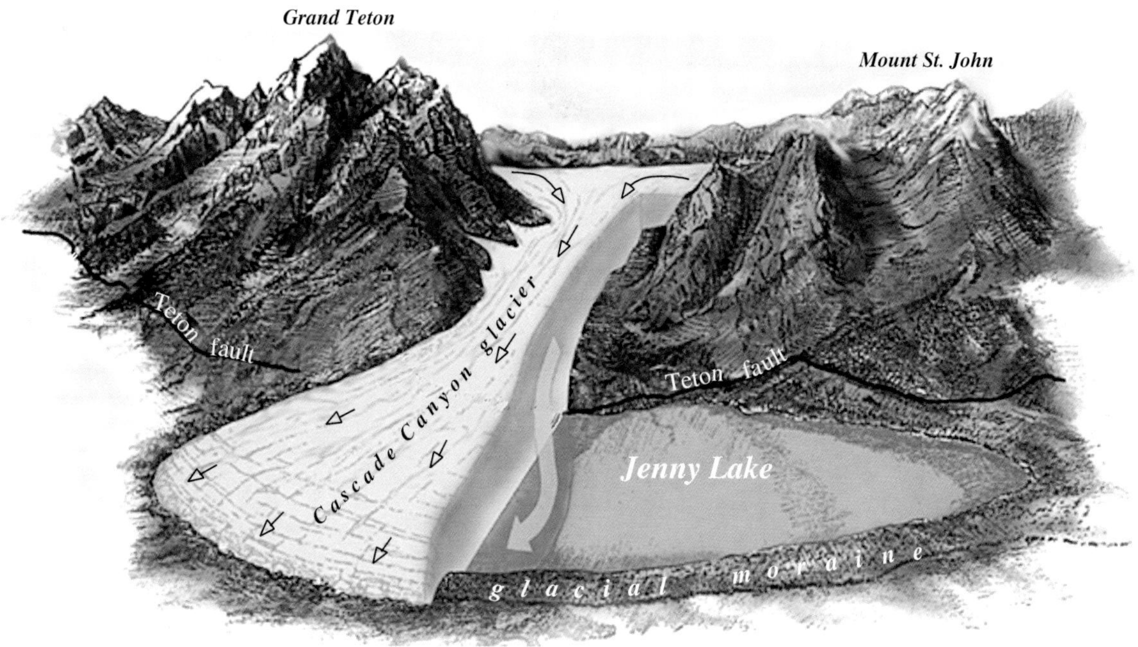

6.5 ~ *Cascade Canyon Glacier is an example of an alpine glacier that flowed east out of the Teton Range. It scoured out Cascade Canyon and deposited a moraine of debris at its base, damming a creek to form Jenny Lake. (National Park Service.)*

Ice fields in the Teton Range fed glaciers that flowed down Cascade, Avalanche, Death, and other major canyons (Figure 6.5), then spread across the floor of Jackson Hole, merging with ice flowing south from the Yellowstone Plateau.

No one knows why the glacial period finally ended, but it likely happened when global temperatures rose 9 to 12 degrees Fahrenheit and precipitation diminished substantially. Pinedale stage glaciers retreated from Jackson Hole some 14,000 years ago, but probably persisted until 12,000 years ago in the Tetons and nearby ranges.

Today, about a dozen small glaciers exist in high, shaded gullies of the Teton Range, but they are not remnants of the Pinedale glaciation (Figure 6.6). They are thought to have formed in what is known as the Little Ice Age, a period from roughly the 1400s through the 1800s. Global temperatures cooled an average of a few degrees Fahrenheit during the Little Ice Age, creating intermittent cold spells in the West and certain other regions.

The largest of the modern glaciers are the Teton and Middle Teton glaciers—located east and southwest of Grand Teton, respectively—and the Falling Ice, Skillet,

6.6 ~ *Schoolroom Glacier, located at the upper end of Cascade Canyon's South Fork, is a classic example for teaching about active glaciers. When this central Teton Range glacier was larger, it scoured rocks from the ground, then deposited them around its edges, forming moraines. A lake formed behind the ridges of debris. (Edmund Williams.)*

and Triple glaciers on Mount Moran. The status of most Teton glaciers has not been tracked. However, Teton Glacier retreated as much as 600 feet between when it was mapped in 1929 and 1963. There is evidence that the Teton Glacier started advancing again in the early 1970s, probably due to increased precipitation.

Rivers Run through It

As climate warmed and the glaciers retreated, they released floods of meltwater that helped carve the near-final forms of the Snake, Yellowstone, and other rivers in the Yellowstone–Teton region. Imagine ice-fed rivers becoming torrents, punctuated by catastrophic floods as ice or debris dams broke. Envision the Grand Canyon of the

Yellowstone churning with raging water, or huge volumes of rock-and-mud-filled water ripping down the Snake, Madison, and other river canyons. These scenes can be compared with devastating floods in Himalayan valleys in recent decades.

In broad valleys like Jackson Hole, streams of melting glacier water formed meanders and braids across what is called an outwash plain. The streams spread glacial debris to create a smooth surface of gravel and sand. Much of the sagebrush-covered floor of Jackson Hole is glacial outwash plain. Such plains are pocked by "kettles," which are pothole-like depressions that formed when water ponded beneath melting glaciers. The plains also are marked by "knobs" or hills of debris dropped from glacial ice. This "pothole terrain" is apparent between the southeast shore of Jackson Lake and the Snake River (Figure 6.7).

6.7 ~ *Grand Teton looms over northern Jackson Hole, where flat, treeless glacial outwash plain* (left half of photo) *was produced by meltwater flowing from a glacier that helped carve Jackson Lake* (right). *The lake was dammed by a glacial moraine, now covered by trees, stretching across the end of the lake. Kettles or potholes* (bumpy areas on the left) *formed where large chunks of ice melted.*

Retreating ice also dumped larger debris. Glaciers ripped old rocks from the highlands of Yellowstone and from the Teton and Gros Ventre ranges. Then ice and meltwaters deposited the debris in valleys such as Jackson Hole and the Madison Valley near West Yellowstone, Montana. In the Lamar River Valley in northern Yellowstone, the sudden failure of an ice dam released a tumultuous flood of water, mud, and rocks that scoured the valley. In some places in the region, glaciers plucked boulders the size of automobiles from the ground and carried them great distances. For example, a boulder of Precambrian granite from the Beartooth Range was carried 15 miles southwest and deposited on the south rim of the Grand Canyon of the Yellowstone.

Major flooding on the Snake cut into the outwash plain, forming a channel with high banks. Then, perhaps centuries later, another disastrous flood came, then another, each cutting the channel deeper and narrower, forming terraces—a series of old river banks rising like stairsteps on each side of the Snake. In the northern half of Jackson Hole, such floods cut almost 140 feet into the valley floor, creating terraces (Figure 6.8). The Snake River Overlook on Jackson Hole Highway provides a good view of these terraces. Terraces also can be seen along abandoned river channels where the Snake at various times flowed from the south and southeast corners of Jackson Lake. One such channel stretches from Spalding Bay southwest to Jenny Lake; the other extends from Spalding Bay southeast to the modern channel of the Snake.

Birth of the Big Lakes

Yellowstone and Jackson lakes—the large, shimmering jewels of Yellowstone and Grand Teton parks—are surrounded by high mountains. The Yellowstone River feeds and drains Yellowstone Lake. The Snake River flows into and out of Jackson Lake. Both rivers originate on the Yellowstone Plateau, near where glacial ice once was thickest. The glaciers carved both lake basins into their present shapes, and also dumped debris to dam the south end of Jackson Lake, forcing the Snake River to exit the lake's east side.

Other processes also helped form both lakes. Jackson Lake may have existed in ancestral form as long as 13 million years ago when the Teton fault became active and a trough started sinking along the fault. Lesser faults and volcanism also helped shape Yellowstone Lake, which may have formed as long as 2 million years ago—after the first caldera eruption blew a big hole in the ground to create a natural lake basin.

Researchers have studied the lakes by bouncing sound waves through the lake beds and drilling core samples of sediments.

6.8 ~ Ancient river terraces flank the Snake River as it flows south through Jackson Hole. As the river cut downward, it left the terraces as remnants of older, higher riverbeds.

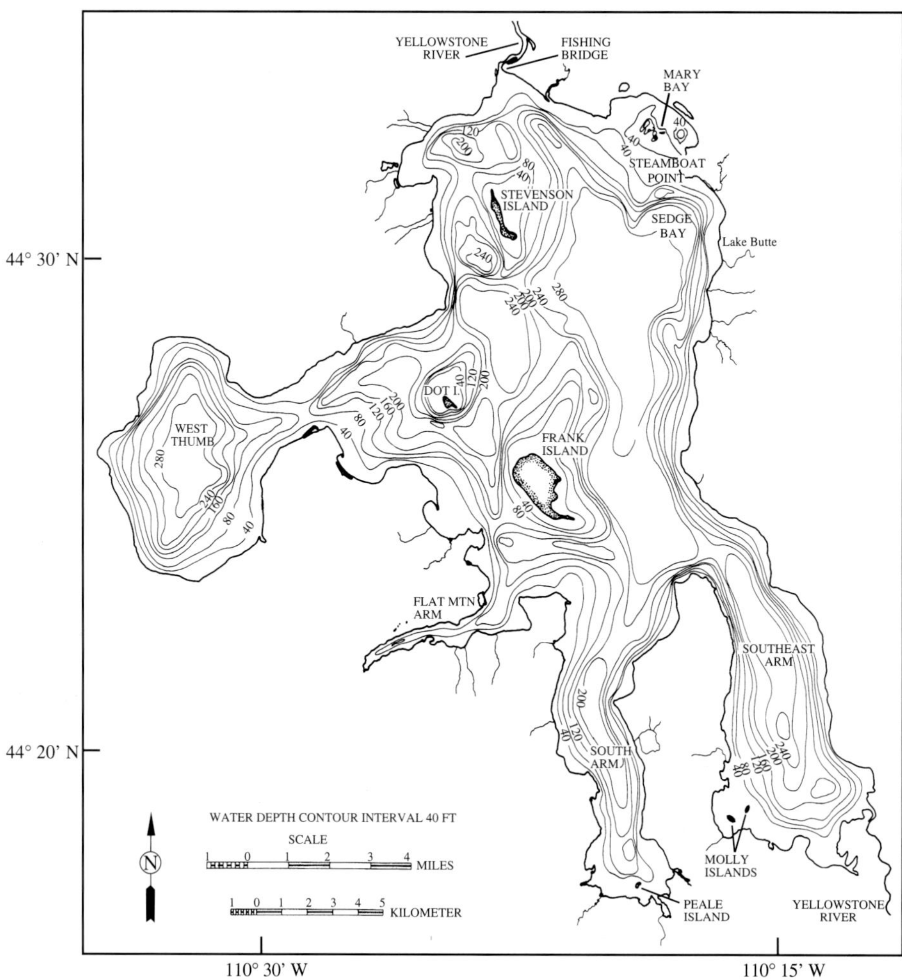

6.9 ~ *Depth map of Yellowstone Lake (contoured in feet below the surface). The deepest part of the lake is on the east side. A smaller, shallower basin at West Thumb was created by a volcanic explosion. Glaciers helped carve and smooth the lake basin and its arms. (Otis, 1977.)*

On a map, Yellowstone Lake looks like a big octopus with a large body and three tentacle-like bays: West Thumb, South Arm, and Southeast Arm (Figure 6.9). The lake is up to 20 miles long from its north shore to the end of either the South Arm or Southeast Arm. It measures about 13 miles from its east shore west through the West Thumb. The central basin of the lake, excluding the arms, measures about 12 miles north–south and 9 miles east–west.

The bottom of Yellowstone Lake has three basins. The main basin, which is as much as 325 feet deep, spans the east side of the lake, running from the Southeast Arm northward to the lake's north end. A second basin, which occupies the South Arm, is about 200 feet deep. A third basin, which is circular, rests beneath West Thumb. It is 280 feet deep, but is separated from the main lake basin by a 60-foot-deep channel.

Volcanism, earthquake faults, and glaciers all contributed to Yellowstone Lake's shape. If Teton-like mountains once occupied the Yellowstone region before the first caldera explosion, lakes similar to Jackson Lake may have sat at the base of such mountains. However, the Yellowstone Lake basin started resembling its current form after the first huge eruption 2 million years ago. The caldera eruption 630,000 years ago would have destroyed parts of the old lake and created a new basin in which modern Yellowstone Lake formed.

The lake's South Arm and Southeast Arm parallel faults that likely helped create troughs occupied by the two bays. The third bay, West Thumb, probably began as a crater from a lava eruption after the last big caldera explosion.

Huffing and puffing of the caldera floor also has influenced the lake's size, shape, and levels over time. Uplift of the caldera floor acted like a natural dam. This dam is not located at Fishing Bridge, which generally is considered the lake's north end and outlet, but 2 miles farther north at Le Hardy Rapids. Upward movement of the caldera floor created a stairstep series of faults that cross the Yellowstone River to create the rapids. Ground on the Yellowstone Lake side of the faults heaved upward while ground downstream dropped downward, helping contain the lake waters.

Between the three major periods when a huge icecap covered all of Yellowstone, it is likely that remnant ice in Hayden Valley—north of Yellowstone Lake and Lee Hardy Rapids—acted as a giant dam to raise the lake level much higher than it is today. Erosion of rock at the north end of the lake may have lowered the lake at times.

Despite the influence of volcanism and faults, Yellowstone Lake's modern shape was largely sculpted by the thick glaciers that repeatedly covered the region and scoured and scooped out the landscape until about 14,000 years ago. While the South and Southeast arms probably started as troughs along faults, they undoubtedly were excavated into broader bays by glaciers.

Jackson Lake also was carved by the glaciers, but faulting played a major role too, explaining the lake's shape and greater depth (Figure 6.10).

Jackson Lake's main, western trough stretches more than 16 miles north to south at the east base of the Teton Range. The main trough is about 3 miles wide. The smaller, eastern trough is about 6 miles long and runs northeast–southwest. The lake

6.10 ~ *Aerial view southwest toward Jackson Lake and the northern Teton Range. The Teton fault runs along the base of the mountains a quarter mile from the lake's west shore. The lake formed because the valley floor dropped down along the fault and tilted west. Glaciers helped excavate the lake basin.*

is more than 7 miles wide at its south end, where the main and eastern troughs join. Jackson Lake is smaller than Yellowstone Lake, but it attains a depth of 437 feet in its western trough, compared with Yellowstone Lake's maximum depth of 325 feet.

The smaller eastern trough was excavated during the Pinedale glaciation by a lobe of ice that flowed southwest into Jackson Hole from what is now Pacific Creek. This trough is only 142 feet deep.

The lake's 437-foot-deep western trough (Figure 6.11) is deeper because it initially was formed by the sinking of Jackson Hole along the Teton fault. Then a lobe of ice flowed from Yellowstone down the Snake River and into Jackson Hole, where it scooped out and enhanced the western trough, making it 800 feet deep. Since the ice receded, the trough has been nearly half-filled by sediments from glacial outwash and subsequent erosion in southern Yellowstone and the Teton Range.

Late in the Pinedale glacial stage, the Snake River and Pacific Creek lobes of ice—along with ice moving east from the Teton Range—all flowed into northern Jackson Hole. The glaciers did not flow as far south and cover the valley as glaciers did during the earlier glacial stages. As the Pinedale ice retreated, it dumped a terminal moraine that dammed the south end of Jackson Lake. That prevented the Snake River from

6.11 ～ *Depth map of Jackson Lake (contoured in feet below the surface). The deepest part of the lake is the western trough, which was created by combined effects of the Teton fault and glaciers. The lake's shallower eastern trough was excavated mainly by glaciers. (Smith, 1993.)*

6.12 ~ *A dam made of glacial moraines wraps around the three sides of Jenny Lake away from the Teton Range. The east shore is the terminal moraine of debris dumped at the toe of Cascade Canyon Glacier, which once flowed east from the mountains and receded 14,000 years ago.*

flowing out of various channels at the south end of the lake, forcing the river to exit at the lake's east side, as it still does today. This moraine, covered with trees, rises 100 to 300 feet above the water level at the south end of the lake.

Moraines also form other distinctive, hummocky, forested ridges in Jackson Hole, including the islands on the east half of Jackson Lake. Other moraines dammed streams to deepen or form lakes such as Taggart, Leigh and Jenny (Figure 6.12). Lateral moraines in Jackson Hole form long ridges, parallel to the direction of ice flow,

6.13 ~ *Phelps Lake, near the south end of Grand Teton National Park, sits within a basin formed by moraines. The moraines were deposited by a glacier that once flowed from Death Canyon, in shadow behind the lake.*

such as Timbered Island, which is just southeast of Jenny Lake, and Burned Ridge, which stretches southeast from Spalding Bay at the south end of Jackson Lake.

The moraines of Jackson Hole contain silts and clays. These soils retain moisture better than looser sediments elsewhere on the valley floor, which explains why forests cover so many moraines, including the aptly named Timbered Island. The tree-covered moraines show where the glaciers halted (Figure 6.13). The glaciers not only chiseled the terrain, but contributed to the vegetation that graces Jackson Hole's landscape.

Future Disasters

7

In 1870, the fall before Ferdinand Hayden's celebrated exploration of Yellowstone, an Army lieutenant named Gustavus C. Doane guided a small troop into the mysterious high country. Unlike Hayden, Doane did not conduct extensive scientific studies. However, Doane was observant. He said of Yellowstone:

> As a country for sight seers, it is without parallel. As a field for scientific research it promises great results, in the branches of Geology, Mineralogy, Botany, Zoology, and Ornithology. It is probably the greatest laboratory that nature furnishes on the surface of the globe.

Yellowstone's value as a unique ecological region soon gained recognition when in 1872, it was designated as the first national park in the United States—and in the world. The complex relationships among Yellowstone's fauna, flora, and geology helped inspire America's budding conservation ethic, which came to fruition only a century later with widespread recognition of the tenuous interdependence of living organisms and the Earth they occupy. The idea of a greater Yellowstone ecosystem recognized that its living and geological wonders extended beyond the park's boundaries and into a broader area. The greater Yellowstone ecosystem is defined by the subterranean yet dominant presence of the Yellowstone hotspot, the engine that ultimately drives not only the region's geology, but also its living organisms.

The Rocky Mountains, lifted upward tens of millions of years ago, were pushed perhaps 1,700 feet higher at Yellowstone during the past 2 million years by the upward-bulging hotspot. Today, a line drawn at 6,100 feet elevation roughly demarcates the boundaries of the greater Yellowstone ecosystem. The high altitude is critical in creating the temperature and moisture regimes that gave rise to Yellowstone's biological wonders and now determine the distribution of its plants and wildlife. In addition, the incredible amount of heat rising from the hotspot is responsible for Yellowstone's history of volcanism and its geysers and hot springs, rich with exotic microbes that branched off the evolutionary tree at a primitive stage of life on Earth.

Yelllowstone's expansive lodgepole pine forests demonstrate the interaction of the park's biology and geology. They grow well on rhyolite lava flows that cover most of western and central Yellowstone. These volcanic soils are not rich, but lodgepoles can survive on them while many other trees cannot. Different plants grow on soils derived from different kinds of underlying rock. Ecologists have noted Yellowstone's grizzly bears often feed on vegetation that grows atop certain rock formations, so their foraging patterns are somewhat predictable by studying geologic maps.

Heat flowing from the ground affects Yellowstone's fish. In icy Yellowstone Lake, lake bottom hot springs foster plant growth and provide havens—nicknamed "trout Jacuzzis"—that attract native cutthroat trout. Fish also can be harmed by Yellowstone's abundant heat. Water from geysers and hot springs sometimes raise temperatures in the Firehole River enough to kill fish. Yet chemicals released from volcanic rocks add nutrients that foster plant growth in the Firehole, providing more food for fish and helping to make the river a blue-ribbon fly-fishing stream.

The hotspot and other geological processes are thus critical factors defining the ecology and life of the Yellowstone country. To some extent, life almost any place on Earth is a product of local geology, or at least dependent on it. For example, after the Rockies were born, the subsequent stretching apart of the Earth's crust created faults like the Teton. Mountains rose and valleys fell along such faults, allowing weather to erode the mountains and send sediments into the valleys to foster lush plant life and the fauna that feed on it. At Yellowstone, the presence of the hotspot and the geology it shaped enhances the relationship between the not-always-so-solid Earth and the organisms that occupy it. It really is the greater Yellowstone "geoecosystem." (See Figure 1.3.)

So it seems cruel, perhaps, that death and devastation can result from the same geological forces that gave rise to Yellowstone's flora and fauna—the forces responsible for scenic wonders that now draw millions of people to Yellowstone and the Tetons each year.

Cataclysmic eruptions, ground-ripping earthquakes, gargantuan landslides, and floods—horrible natural disasters shaped the wonders of Yellowstone and the Teton Range long before modern humans evolved and spread around the globe. A human life is but an instant in geologic time, so people have trouble reacting to the threat of infrequent geologic catastrophes. Yet as sure as time proceeds, the same geologic forces that shaped and sculpted this beautiful landscape will do so again—with awful consequences for living creatures.

Nightmare on the Teton Fault

No one can predict exactly what will happen when the Teton fault ruptures again. But a worst-case scenario can be envisioned based on experiences from other major earthquakes and disaster plans prepared by local government.

It could occur in the dark of night during late spring as thousands of visitors slumber in tents, trailers, and cabins throughout Jackson Hole. First comes the low rumbling, which quickly builds in volume and intensity into violent shaking that lasts perhaps 30 seconds or more, although it seems like a lifetime. In canyon campgrounds on the edges of Jackson Hole, a few unfortunate people are crushed to death as the earthquake sends large boulders tumbling off hillsides. In Jackson, Wyoming, some people are tossed from their beds; others jump awake, scramble around, and fall to the ground in panic as windows shatter and bookshelves, refrigerators, and other loose items and furnishings topple. A few brick structures collapse. The number of deaths is low, partly because the town has few multistory buildings.

Unseen in the darkness of Jackson Hole, the ground rolls in waves like the surface of the sea. Landslides along the front of the Teton Range send huge clouds of dust skyward. Slides in the Gros Ventre, Hoback, and Snake river canyons block roads. Avalanches and rockfalls plunge off high Teton peaks, entombing a few scattered groups of mountaineers asleep in their tents.

For some 35 to 40 miles along the east base of the Teton Range, the earth ruptures during the violent shaking. On one side of the fault, the mountains jerk upward; on the other side, Jackson Hole slips downward. The result is a steep cliff up to 12 feet tall that didn't exist moments earlier. The length and height of the cliff indicate this was a Big One on the Teton fault, a quake measuring perhaps 7.5 in magnitude. Some shaking is felt throughout the West, as far away as California, Oregon, and Washington.

Throughout Jackson and Jackson Hole, families are huddled outside homes and cabins, worried about reentering because of structural damage. Many more are left with broken water, fuel and sewage or septic lines. Power has failed throughout the valley, and in some spots downed lines set off brush or forest fires. In nearby Bridger–Teton National Forest, several campers are stranded because slides or fallen trees blocked roads.

To the north, in Yellowstone, there are more landslides, blocking narrow canyons and crushing several campers. Some geysers cease, others erupt for the first time as the shaking breaks up mineral deposits clogging hot water conduits underground.

Landslides on the margins of Jackson Hole have blocked streams such as Cache Creek, the Gros Ventre River, Flat Creek, and Buffalo Fork, creating earthen dams that might fail within days and flood downstream areas. Flooding is also likely where levees break along the Snake River above Wilson, Wyoming. The worst potential flood threat lurks upstream at Jackson Lake Dam, which once was vulnerable to collapse during a moderate, magnitude-5.5 quake. The U.S. Bureau of Reclamation strengthened the dam during 1986–1989, promising it would withstand the maximum credible earthquake on the Teton fault, a magnitude-7.5 jolt centered 4 miles away.

7.1 ⁓ *U-shaped glacial valleys descend east from the Teton Range, which was uplifted during large prehistoric earthquakes on the Teton fault.*

For the sake of envisioning a worst-case quake on the Teton fault, imagine what would happen if Jackson Lake Dam did fail, spelling disaster for an area extending at least 80 miles down the Snake River to Palisades Dam. If Jackson Lake Dam collapsed during heavy spring runoffs and Palisades already was nearly full, Palisades would have to release large volumes of water from its spillways, possibly flooding agricultural land farther downstream along the Snake River Plain in Idaho.

The worst, however, would happen upstream. As the quake rocked Grand Teton National Park, 6-foot-high waves would slosh across Jackson Lake. As the dam collapsed, water would gush outward, sucking in any boats moored within a half-mile of the dam. The water would race eastward over Oxbow Bend. Within a half-hour it would engulf the Buffalo Ranger Station and the park's Moran entrance station. If an alarm system worked as planned, sleeping Park Service employees might escape with their lives.

The flood would move surprisingly slowly. Five hours after the quake and almost 25 miles downstream from the breached dam, the surge would wipe out Grand Teton National Park headquarters at Moose, empty for hours thanks to an alarm system that warned resident park and concession workers. Next to be inundated would be homes in the Snake floodplain south of Moose. A few hours later, Teton Village would be cut off from the outside world as the flood covered roads. Next, the town of Wilson and homes in surrounding areas would be flooded.

The town of Jackson, away from the river, would be spared the insult of flooding in addition to shaking damage. But flooding south of the town and quake damage on the north could destroy bridges at both ends of Jackson, isolating the city from the outside world.

It would take almost 17 hours for the flood to reach Palisades, and as it roared down the Snake River Canyon, it would claim the lives of several more boaters who were stranded by landslides or not warned of the deluge.

Dozens of people probably would die during a major quake on the Teton fault. Hundreds would suffer injuries. If Jackson Lake Dam failed, the toll could be much higher.

Betting on Seismic Disaster

What are the chances of another magnitude-7 or stronger quake on the Teton fault? Estimates outlined in Chapter 5 suggest a major quake happens once every 680 to 4,000 years. To avoid undue precision, let's just say once every several hundred to several thousand years. The most recent quake occurred between 4,840 and 7,090 years ago. So a major quake seems overdue.

It is possible, probably even likely, that the rates of activity on the Teton fault have changed over time. Evidence indicates it may be less active now than it was for several thousand years beginning 14,000 years ago, the time the glaciers receded. Perhaps the ground rebounded as the heavy weight of ice was removed, making quakes more frequent during that period. Today, Earth's crust continues to stretch apart in the region, indicating the fault remains active and is accumulating stress that eventually will produce another magnitude-7 or stronger quake.

The Teton fault is one of at least five faults capable of generating a major quake in the greater Yellowstone–Teton region. Such quakes also could occur on any of several known faults on the periphery of the Yellowstone Plateau. They include the Hebgen Lake fault, responsible for the 1959 disaster; the Red Mountain fault at the south end of Yellowstone National Park near Mount Sheridan; the South Arm fault bordering the west side of Yellowstone Lake's South Arm; and the East Gallatin fault in northwest Yellowstone.

Including the 1959 Hebgen Lake quake and the last two prehistoric jolts on the Teton fault, we have direct evidence of at least thirteen major quakes within the last

25,000 years on the five significant faults. This suggests a magnitude-7 quake happens somewhere in the region once roughly every 1,900 years. However, there is less direct evidence that the Teton fault alone likely produced at least thirteen big tremors in the last 14,000 years. The other faults may be similarly active. So it is not unreasonable to guess perhaps 100 major quakes have rocked the area in the past 25,000 years, or once every 250 years. That would make the annual probability of another magnitude-7 disaster 1-in-250. On the other hand, it could be argued a major quake is much less likely any time soon because two major quakes already struck the Intermountain West in recent decades: the magnitude-7.5 Hebgen Lake quake and the magnitude-7.3 Borah Peak quake of 1983 on the Lost River fault.

In addition to the area's major faults, numerous smaller faults within the Yellowstone caldera can generate quakes, probably as large as magnitude 6.5, plenty strong to cause havoc and damage some older buildings within the national park.

Somewhat farther from the Yellowstone–Teton area—but still within strong shaking distance—are numerous other faults capable of producing major quakes in the range of magnitude 7 to 7.5. They include the Lost River, Lemhi, and Beaverhead faults in Idaho and the Hoback, Star Valley, Grand, and Greys River faults in western Wyoming south of the Tetons. All these faults are located on the edges of the "wake" left by the Yellowstone hotspot as Idaho slid over it. Many of the western Wyoming faults are considered overdue for a major jolt.

The risk of earthquakes also increases the already high danger of landslides. Wyoming officials mapped the state's landslide threat and determined two of the state's highest landslide hazards were in the Snake River drainage, including most of Teton County, and in the Upper Yellowstone–Madison drainage basin, which includes much of Yellowstone National Park.

What if the annual chance for a major quake ripping apart the Teton fault is 1-in-250, 1-in-2,500, or 1-in-4,000? All those numbers seem low. Are they worth worrying about? Consider the odds of more familiar events and tragedies to gain perspective on disasters that happen in geologic time. Although estimates vary, some commonly cited odds are that each year, an American has a 1-in-6,000 chance of dying in a car wreck, a 1-in-10,000 chance of death by homicide, a 1-in-1-million chance of dying in the bathtub, a 1-in-2-million chance of being killed by lightning, a 1-in-3-million chance of freezing to death and a 1-in-10-million chance of dying in a commercial airplane crash.

If the odds of a major quake or volcanic eruption seem too small to raise concern, consider the millions of people who enter multistate lotteries like Powerball, even though the chances of hitting the jackpot can be as low as one in tens of millions.

Many people equate a low chance of a geologic disaster as no chance at all. This implies the event never will happen—an incorrect belief. After all, someone always wins the lottery, even if any individual's chance of winning is tiny. So although most people have a minute chance of experiencing or dying in a major quake or eruption, many people die in quakes each year, and deadly volcanic eruptions are not uncommon. The consequences of such catastrophic events—when they do occur—can be more devastating than more common disasters such as floods, tornadoes, and hurricanes. Yet floods and storms are more common in our lifetimes. We tend to disregard the risk of less common volcanic and seismic disasters. Nevertheless, it is wise to prepare for them when living in or visiting a geologically active area like the Yellowstone–Teton region.

When Will Yellowstone Erupt Again?

Future volcanic eruptions can be expected at or near Yellowstone for the simple reason there is hot and molten rock, or magma, beneath the caldera now.

There is no evidence the crust of North America is stopping or changing its southwest movement in a way that would terminate the stretching of Earth's crust and the ascent of heat and molten rock from the Yellowstone hotspot. Even if the supply of heat from deep within Earth was stopped somehow, it could take thousands of years for molten rock already in Earth's mantle and crust to stop rising. For now, it is likely the hotspot continues to heat molten rock beneath Yellowstone.

Yellowstone's 2-million-year volcanic history, the amount of heat that still flows from the ground, the frequent earthquakes, and the rising and falling of the caldera floor also testify to the likelihood of continued volcanism. Calderas such as Yellowstone are known as giant volcanoes at unrest because of their large size and characteristic huffing and puffing. They are like snoring beasts, and no one knows when they will awake again.

Just as large earthquakes occur less often than small quakes, cataclysmic volcanic eruptions are infrequent compared with lesser eruptions. Such has been the case during the 2 million years since the hotspot started generating volcanism at Yellowstone.

Yellowstone's volcanic history suggests the potential for various kinds of eruptions in the future. Most likely would be steam and hot-water explosions from shallow reservoirs of hot water—not molten rock. Such explosions could blast out craters more than a mile wide, such as those in the northern Yellowstone Lake Basin, including the lake's Mary Bay and nearby Indian Pond. Both were produced by prehistoric steam blasts within the past several thousand years.

The next most likely kind of volcanism at Yellowstone would be lava flows produced either by explosive eruptions of lighter rhyolite lava or gentler eruptions of darker basalt. Such eruptions could generate lava and ash roughly comparable in volume to the 1980 eruption of Mount St. Helens, although Yellowstone eruptions likely would yield more lava and less airborne ash. Rhyolite or basalt eruptions within the Yellowstone caldera also could be much larger, similar in scale to the 1991 Mount Pinatubo eruptions in the Philippines and releasing at least ten times more material than the Mount St. Helens blast. The lava could spread across the surface and create a layer tens of feet thick. Rhyolite eruptions would resemble those that filled the Yellowstone caldera since it last exploded about 630,000 years ago. Basalt flows would resemble those that covered the Snake River Plain in the wake of the Yellowstone hotspot.

The least likely but worst-case volcanic eruption at Yellowstone would be a gigantic caldera explosion such as those 2 million, 1.3 million, and 630,000 years ago. Those are only the most recent in the swath of calderas extending from southwest Idaho to Yellowstone. These calderas testify to the repeated cycles of caldera volcanism generated as North America drifted southwest over the Yellowstone hotspot. During the past 16.5 million years, each cycle began with a colossal explosion that blew a hole in the Earth measuring tens of miles wide. Such explosions were followed by smaller, but still significant eruptions of light-colored rhyolite lava. The next stage featured eruptions of darker, basaltic lavas like those that now hide the ancient calderas on the Snake River Plain.

The nature of the next eruption at Yellowstone depends on whether we are at the end of the cycle that began with the massive caldera blowout 630,000 years ago or at the beginning of a new cycle. We simply don't know the answer to that question. If we are at the end of the cycle, relatively gentle eruptions of basalt lava are most likely—or perhaps an end to volcanism in the area. If Yellowstone is near the start of a new cycle, that would imply we are nearing the time of another caldera catastrophe. There is some evidence suggesting Yellowstone might be nearing a new explosive cycle within 100,000 years: Concentrations of the element neodymium are highest in some of Yellowstone's youngest rocks of rhyolite lava. Such high values are associated with large caldera-forming eruptions.

Pompeii, USA

Almost 12 million years ago, a marshy watering hole in eastern Nebraska attracted hundreds of animals from the surrounding plains. Horses, camels, rhinos, cranes and other birds, giant tortoises, sabertooth deer, and fox- and raccoon-like creatures gathered near the water. Then, one day, the sky turned dark as a vast cloud of volcanic ash swept in from the west. Visibility fell to zero, and there was no place to hide as the gray dust fell onto the plains. Within days, a foot-deep layer of ash covered much of the ancient Great Plains. Birds and small predators probably died soon after the ash cloud hit. It took longer for the panicked horses, camels, and rhinos to choke to death on the ash and die. As weeks progressed, winds blew ash from the hills and into low spots, including the watering hole, where the gritty gray powder accumulated in 8-foot-deep drifts, covering and preserving the carcasses of hundreds of creatures, including mother rhinos with their babies right next to them.

Millions of years later, Nebraska's modern human inhabitants began to discover the bones of the animals killed by volcanic ash when the Yellowstone hotspot produced a giant caldera eruption roughly 800 miles away in southern Idaho's Snake River Plain. The Ashfall Fossil Beds State Historical Park near Royal, Nebraska, is a prairie Pompeii, a scene frozen in time like the Italian city buried by pyroclastic flows—flows of hot boulders, rocks, ash, and gas—from the A.D. 79 eruption of Vesuvius. This small section of Nebraska also gives a hint of the magnitude of disaster that will occur when a Yellowstone hotspot caldera explodes again. The eruption in southern Idaho 12 million years ago, like the Yellowstone caldera blast 2 million years ago, sent thick clouds of ash over much of what is now the western half of the United States, including California. The caldera explosions at Yellowstone 1.3 million and 630,000 years ago were somewhat less apocalyptic.

If the Yellowstone caldera blew up today, cattle, sheep, and wild animals living outdoors across the West and Midwest would die like the horses, rhinos, and camels in prehistoric Nebraska. People could seek shelter in their homes and other buildings. Nevertheless, activities by millions of people would be difficult for weeks or longer until rain helped pack down and wash away inches to feet of ash covering a third to a half the nation (Figure 7.2). Many people likely would die.

Consider, for example, the 1991 eruptions of Mount Pinatubo in the Philippines. Pyroclastic flows of hot rock and other debris buried the landscape many yards thick extending 6 miles or more from the volcano. Mudflows streamed down valleys, burying thousands of homes. An area extending 90 miles from all sides of Pinatubo was blanketed by ash and heavier rocks and blocks of pumice. Ash was 20 inches deep as far as 50 miles from the volcano. Huge ash plumes blocked the sun in Manila, 60 miles away. Storms combined with the eruptions to make the volcanic ash wet and thus heavier than normal, causing numerous buildings to collapse. By the time the eruptions ended, some 300,000 people had been evacuated and at least 700 had died from

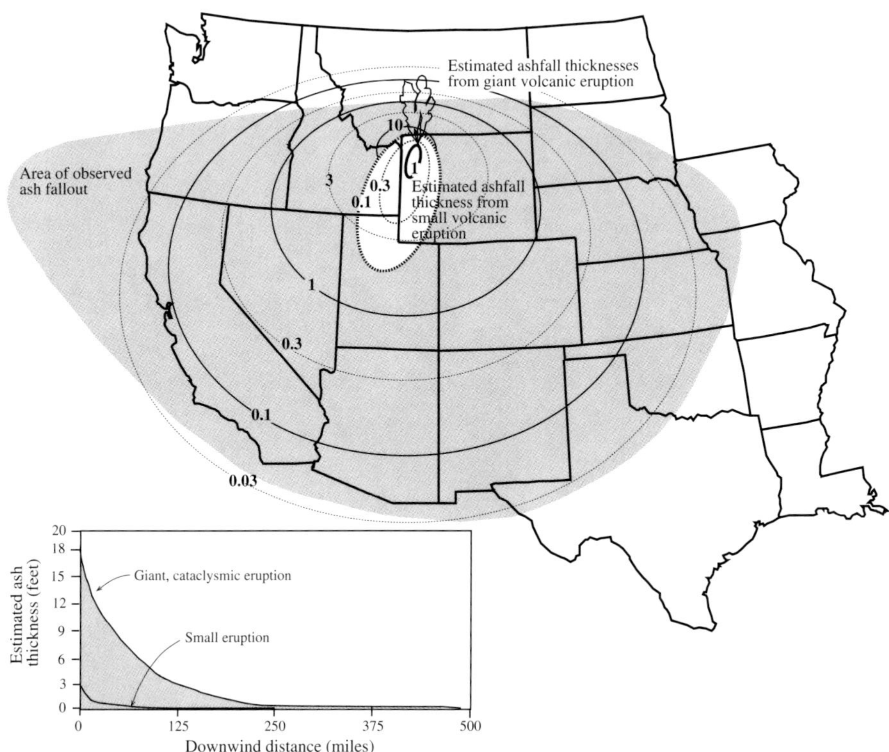

7.2 ~ *Depths of volcanic ash that could be deposited by future caldera eruptions (gray) and by smaller eruptions (white oval) at Yellowstone. Prevailing winds would determine actual ashfall patterns. Contour lines show ash depths in feet. (Michael Perkins.)*

the effects of the eruptions. More than 320 were killed in the climactic eruption of June 15, 1991, mostly when they were crushed by the collapses of ash-covered roofs. The total death toll included tribesmen who contracted deadly diseases in refugee camps after being displaced by the eruptions.

Gases and particles from Pinatubo helped deplete Earth's protective ozone layer, created colorful sunsets, and cooled Earth's climate slightly for a couple of years, briefly moderating a global warming trend. The eruptions sent a 3,000-mile-wide belt of gas and dust around the planet.

Yellowstone caldera eruptions 2 million, 1.3 million, and 630,000 years ago were about 250, 28, and 100 times larger than the 1991 Pinatubo eruptions, respectively. If such an eruption took place today, the consequences on the environment and human life would be extraordinary, far beyond those from any catastrophe in modern time.

It is likely a gargantuan caldera eruption at Yellowstone would be preceded by months to years of earthquakes and by the uplift of the caldera floor a few yards and, possibly, by some small basalt lava flows. But calderas around the world often huff and puff for decades without producing cataclysmic eruptions. We can only guess whether the geologic warnings would be adequate to prompt the evacuation of Yellowstone and surrounding areas and towns to prevent the instantaneous loss of thousands and perhaps tens of thousands of lives.

Devastation would be complete and incomprehensible at the caldera. Imagine Yellowstone National Park and everything in it destroyed. Every road, every lodge, every campground, every visitor center, every geyser and scenic feature would either be blown instantly off the face of the Earth or swallowed as the floor of the caldera sank downward during the eruption.

The initial blast would be followed by extensive pyroclastic flows, huge mudflows thundering down river valleys, and volcanic ash blown high into the atmosphere and falling in a layer feet to inches thick over the western half of the United States. Such an eruption could expel hundreds to thousands of cubic miles of lava, ash, and other debris.

Nearby towns like West Yellowstone, Gardiner, and Cooke City, Montana, likely would be wiped out by ash, mudflows, or pyroclastic flows. Life probably would come to a standstill for weeks or months as several feet of ash buried cities somewhat farther away like Cody and Jackson, Wyoming; Bozeman, Montana; and Idaho Falls, Idaho.

A Yellowstone caldera explosion today would dump ash over hundreds of thousands of square miles, not the mere thousands of square miles buried by Pinatubo. Ashfall could reach 20 feet near the blown-out caldera and, depending on winds, perhaps a few feet of ash would cover Salt Lake City, almost 300 miles away, and a foot could fall on Denver, more than 400 miles away. That may seem amazing, but a Yellowstone caldera eruption would be orders of magnitude larger than the 1980 Mount St. Helens blast or the catastrophic eruption of Oregon's Mount Mazama 7,600 years ago. Mount St. Helens deposited ash 20 inches thick a few miles from the volcano and sent a thin layer as far as 300 miles. Mount Mazama's blast, which left a hole in the ground now filled by Crater Lake, dumped 5-foot-thick ash within 15 miles of the mountain and spread thinner layers as far as 600 miles.

Even a much smaller, Pinatubo-sized eruption in the Yellowstone caldera would deposit a foot or more of ash within 10 miles of the eruption vent, with a dusting of ash falling as far as 400 miles away, the distance of Denver, and covering much of Wyoming, Montana, Idaho, Utah, and Colorado. Such an eruption also could send lava flows several hundred feet thick over areas of up to 40 square miles within Yellowstone. A Pinatubo-sized eruption at Yellowstone, even though much smaller than the worst-case caldera blast, would make worldwide news and severely disrupt river drainages, roads, and tourism in and around the national park.

Depending on wind conditions, ash from a major caldera explosion could disrupt interstate transport of goods across most of the West and Midwest. Local winds could capture the fine particles and envelop the region in a veil of dust, coating buildings and agricultural lands. By one estimate, the ash could cause global famine because half of the world's cereal grains are produced on North American farmland, much of which would be covered by Yellowstone ash.

Rain would turn the ash to mud that would cover some areas as a gooey layer and flow into rivers, clogging them. For example, the sediment load of the Mississippi River could be increased 400 to 800 times if the Mississippi drainage received only one-fifth of the ash generated by the caldera blast at Yellowstone 2 million years ago. That could increase flooding significantly, impair shipping, and make it impossible for power plants to use silty river water for cooling. Mudflows from Mount St. Helens's 1980 eruption blocked Columbia River shipping channels, creating a months-long disruption of international trade with Portland, Oregon. Imagine what ash from a Yellowstone eruption hundreds to thousands of times larger might do to clog major rivers, trigger flooding, wreck hydroelectric turbines, and disrupt shipping.

A future caldera explosion at Yellowstone could send ash nearly 20 miles high and carry fine particulates and gases around the globe. A Yellowstone cataclysm might reduce global temperatures almost 2 degrees Fahrenheit—cooling that sounds small but is significant enough to be considered a "volcanic winter." After an Icelandic volcano erupted in 1783, Ben Franklin noted lower temperatures during that year and 1784. The 1815 eruption of Tambora in what is now Indonesia had worldwide effects, and was followed by what has been called "The Year without a Summer." In 1816, snow reportedly fell during much of the year in New England. There were widespread crop failures elsewhere. Northern hemisphere temperatures decreased by more than 1 degree Fahrenheit. By any standards, a Yellowstone-sized caldera eruption would be far larger than the Tambora disaster.

We can only speculate on the disaster that society would face during a Yellowstone caldera eruption. As unlikely as such gargantuan explosions seem in our lifetimes, they have happened before and will occur again. We cannot control or stop such overwhelming geological forces, but we can reduce the potential scope of the catastrophe by installing modern surveillance systems to track earthquakes, deformation of Earth's crust, changes in hydrothermal gases and fluids, and other signs of impending eruptions. Such warning systems will allow evacuation of potential areas affected by such eruptions and minimize damage and loss of life.

Volcanic Odds

There are many ways of trying to estimate the volcanic hazard at Yellowstone. No single method may be accurate, yet together they provide a sense of the level of risk.

If the entire West is considered, we know of at least fourteen notable eruptions in the last 2 million years, ranging from the relatively small outbursts at California's Lassen Peak from 1914 through 1917 to the three explosions at the Yellowstone and Island Park calderas. Other peaks that erupted during the last 2 million years are Washington's Mount St. Helens, Oregon's Mount Mazama and Newberry caldera, New Mexico's Valles caldera, and California's Long Valley caldera, where the ski town of Mammoth Lakes is located. Fourteen significant eruptions in the West in 2 million years means such a disaster happens in the region once every 143,000 years. That implies annual odds for a moderate to big eruption somewhere in the West are 1-in-143,000.

A much different answer is reached by considering only the last 8,000 years, during which there were at least seven significant eruptions in the West: Mount Mazama, two at Lassen Peak, and four at Mount St. Helens. Seven eruptions in 7,000 years works out to annual odds of about 1-in-1,100. Of course, none of them were huge caldera explosions.

Now, instead of looking at the West, consider only the gigantic caldera eruptions from the Yellowstone hotspot between the time the hotspot erupted near the Idaho–Oregon–Nevada border 16.5 million years ago to the most recent caldera explosion at Yellowstone 630,000 years ago. During that time, there have been about 100 giant caldera eruptions concentrated in seven to thirteen volcanic centers—each center the site of overlapping caldera blasts. This means eruptions occurred an average of every 160,000 years or so as the hotspot repaved the Snake River Plain and Yellowstone. However, the eruptions became less frequent over time (Figure 7.3), apparently because the supply of magma or heat or both decreased as North America moved over the hotspot.

If we consider the last 6.5 million years, there were six giant caldera-forming eruptions averaging 1.2 million years apart. Since the last blast was 630,000 years ago, this calculation implies the Yellowstone hotspot is unlikely to blast out a new caldera for another 350,000 years.

Yet if we analyze only the three caldera explosions since the hotspot reached Yellowstone—the blasts 2 million, 1.3 million, and 630,000 years ago—the average time between catastrophes is about 685,000 years, which implies another cataclysm is due in 55,000 years.

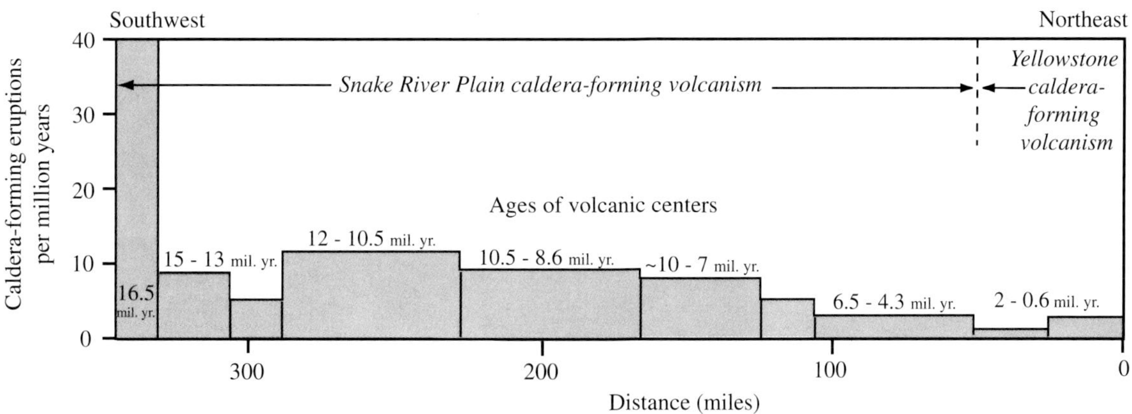

7.3 ⁓ *Approximate dates and number of caldera eruptions per million years from volcanic centers along the Snake River Plain and at Yellowstone. The cataclysmic eruptions have become less frequent over time. (Adapted from Perkins and Nash, in press.)*

Still other estimates of volcanic risk can be reached by focusing on more recent significant volcanism, not just major caldera eruptions. Since the Yellowstone caldera last blew up 630,000 years ago, there have been about thirty lava flow and ash eruptions within the caldera, most recently 70,000 years ago. It might be argued the threat of volcanism at Yellowstone has passed because 70,000 years have elapsed. On the other hand, the average time between each of the thirty eruptions was about 19,000 years. So it also could be argued Yellowstone is more than 50,000 years overdue for another eruption. To further muddy the picture, realize that 19,000 years is the average time between eruptions; the longest interval between eruptions was 148,000 years. So even with the last eruption 70,000 years ago, it is possible to argue the next one is tens of thousands of years in the future.

All these exercises in estimating volcanic risk come up with different answers, but they nevertheless tell us that volcanic eruptions in the Yellowstone region are much less likely than major quakes. The annual risk of a cataclysmic volcanic caldera explosion at Yellowstone is somewhere in the range of one-in-tens-of-thousands to one-in-hundreds-of-thousands or more. This means such catastrophes are extremely unlikely within coming generations and more probable within tens of thousands or hundreds of thousands of years. Annual odds may be one-in-thousands for lesser but still significant eruptions of lava and ash—roughly on the scale of the 1980 Mount St. Helens eruption or the much larger 1991 Mount Pinatubo eruption. The odds are progressively higher for smaller volcanic outbursts, such as steam explosions.

Yet even the risk of a catastrophic eruption at Yellowstone is much larger than the odds of winning a multistate lottery jackpot.

The Future Hotspot

Just as small eruptions are more likely than larger ones, eruptions from existing volcanic vents occur more often than eruptions of brand new volcanoes. An eruptive vent can form only after molten rock has fractured Earth's crust to create a new conduit to the surface.

Where is the Yellowstone hotspot going? The North American plate is moving inexorably about 1 inch per year southwest across the hotspot, leaving in its wake "old" Yellowstones such as the ancient calderas hidden beneath the Snake River Plain. It is possible the hotspot will move northeast over millions of years, blasting a new 40-mile-wide caldera in the Beartooth Mountains or perhaps raising the present site of Billings, Montana, a couple thousand feet and then blowing it to oblivion during a caldera eruption. (Attention Billings residents: There's plenty of time to move.)

In the hotspot's wake, the present site of Yellowstone National Park could lose a couple thousand feet of elevation as it sinks in the wake of the hotspot, then get repaved by the kind of basalt lava flows that covered Idaho's fertile Snake River Plain. If humans still are around in millions of years, the geysers and other glories of today's national park might be replaced by the slogan, "Yellowstone: Famous Potatoes."

On the other hand, the hotspot started blowing huge holes in the ground near today's Oregon–Idaho–Nevada border 16.5 million years ago—a time of accelerated stretching apart of Earth's crust in the Basin and Range Province. Such stretching may well ease the ascent of buoyant masses of hot and molten rock from the hotspot, culminating in new caldera eruptions on the surface, essentially creating new Yellowstones. However, most of the stretched and thinned crust of the Basin and Range Province has passed over the hotspot. Now, as North America's crustal plate moves southwest, it is pulling the thickened crust of the Rocky Mountains over the hotspot. Earth's crust northeast of Yellowstone not only is thicker but is also colder, which means more solid rock must be heated and melted before magma can erupt to the surface. The crust northeast of Yellowstone also is being squeezed—not stretched—giving it greater strength. These factors may make it more difficult and time-consuming for hotspot magma to rise, heat and melt overlying rock, and ultimately trigger volcanic outbursts. Or perhaps the rise of molten rock could stall completely. Only a few million years will tell.

Grand Teton Tour 8

How to Use This Tour Guide

Because winter snows close roads in both Grand Teton and Yellowstone national parks, the driving tours in this chapter and the next are intended for use only from late spring through early fall. You may wish to do only parts of each tour and so we have not shown cumulative trip mileage in these tour guides. Instead, we provide cumulative mileage only from one stop to the next, and for points of interest between them. This chapter's tour of Grand Teton National Park totals 82 miles, excluding mileage to the optional aerial tramway ride.

The intent of these two chapters is to provide a three-day driving tour, including one day in Grand Teton and two in Yellowstone. However, you easily may extend the tour to five days or even longer if you choose a leisurely pace or decide to make optional hikes and stops. The three-day tour outlined in these chapters starts in the town of Jackson, Wyoming. Our tour includes the following suggestions:

- On day 1, make the Teton tour, perhaps beginning or ending with the optional tramway ride detailed at the end of this chapter. Spend the night either in Jack-

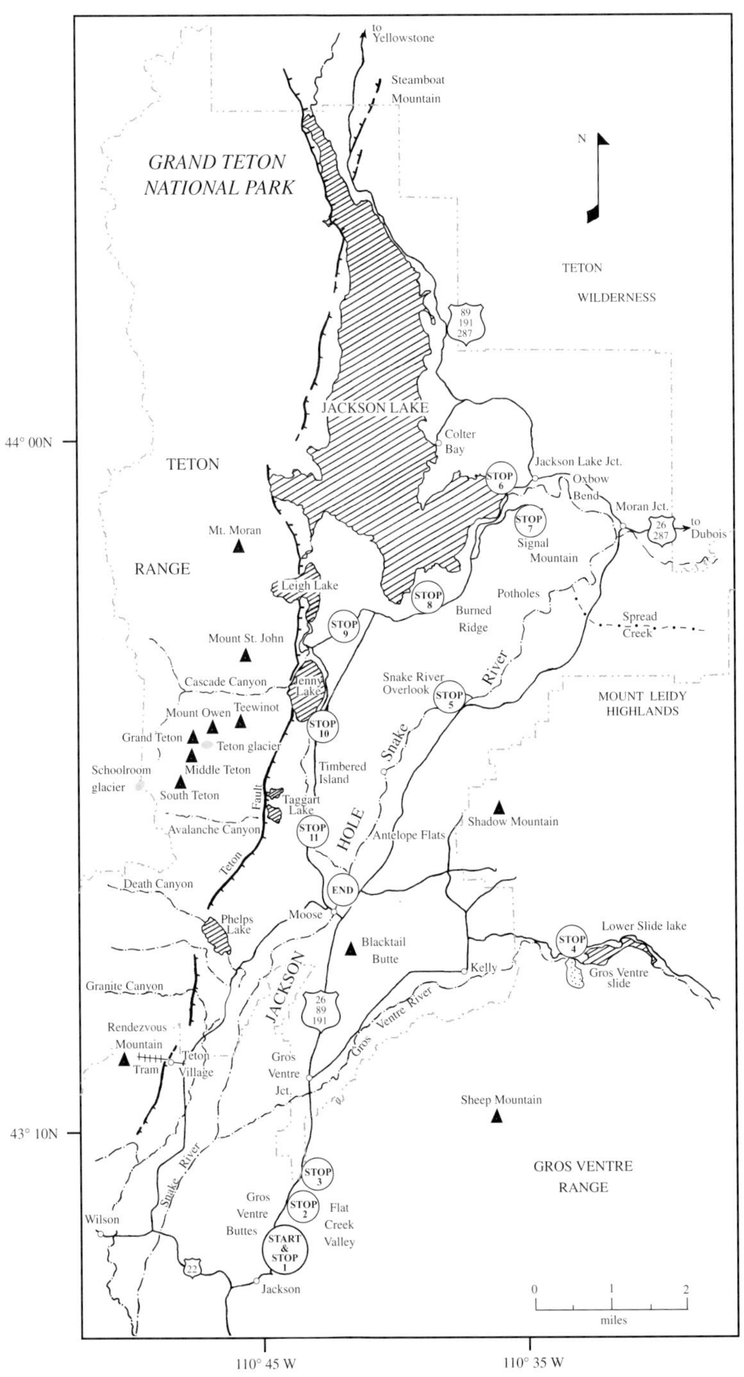

8.1 ⁓ *Index map with numbers for each stop during this chapter's Grand Teton National Park tour, which begins in Jackson and ends in Moose, Wyoming.*

son or find accommodations closer to Yellowstone, such as at Colter Bay Village or other campgrounds and lodgings in northern Grand Teton National Park.

- On day 2, enter Yellowstone's south entrance and drive the loop road clockwise to Madison Junction, then spend the night at West Yellowstone, Montana. If you arrive at West Yellowstone by early to mid-afternoon, you still will have time to make the optional tour to the Hebgen Lake earthquake area, although the visitor center there closes in the late afternoon.

- On day 3, either start with the optional side trip to the Hebgen Lake earthquake area, or proceed from West Yellowstone, Montana, back into Yellowstone National Park, continuing the tour at Madison Junction.

Some visitors may choose to drive part or all of these tours in a direction opposite to the one we use here. For that reason, we also provide reverse mileage between each stop and the sights between stops. However, instructions for left or right turns are written assuming that visitors follow the tours in the direction given.

Vehicle odometers vary, so mileages should be taken as approximate. Primary geological sights in this chapter are labeled Stop 1 through 11. (Figure 8.1. See also Figure 1.4 for a view of the region's topography.) Lesser sights are included in the "On the way to Stop X" sections between the stops.

Some stops are turnouts along the side of a road; others involve driving down an access road or into a large parking lot. In all cases, the stop-to-stop mileages in this tour include distances from one parking location to the next, not simply to the entrances of parking lots.

For background, we suggest reading Chapter 5, The Broken Earth, and Chapter 6, Ice over Fire.

Introduction

As you tour Grand Teton National Park, you will be driving through an intersection of different landscapes shaped by distinctly different geological processes:

To the south and east is the Wyoming Overthrust Belt, a region where rock layers were uplifted, folded, and thrust over one another by east–west squeezing of Earth's crust. This happened during uplift of the Rocky Mountains between 80

8.2 ~ The central Teton Range with the Cathedral Group of peaks, including 13,770-foot Grand Teton, the highest peak, right of center.

and 30 million years ago. The squeezing and uplift made this region a highland long before the Teton fault and modern Teton Range were born.

To the southwest, Earth's crust has been stretching apart for more than 17 million years, creating the Basin and Range Province, a region of alternating north–south mountain ranges and north–south valleys that extends from the Tetons through southern Idaho, southeast Oregon, western Utah, Nevada, and into southern California. This stretching reached the Teton–Jackson Hole area about 13 million years ago, creating the Teton fault. Movement on the fault since has made the modern Teton Range rise upward and Jackson Hole drop downward, separating the Tetons from the Gros Ventre Range.

About 16.5 million years ago, the Yellowstone hotspot started producing catastrophic volcanic eruptions near what is now the Oregon–Idaho–Nevada boundary. As North America drifted southwest, eruptions from the hotspot moved northwest, blasting out and repaving what is now the Snake River Plain in Idaho southwest of here. A few million years ago, the hotspot reached Yellowstone, uplifting what is now the Yellowstone Plateau north of here and producing three gargantuan volcanic eruptions 2 million, 1.3 million, and 630,000 years ago. Glaciers from high atop Yellowstone flowed south and helped shape Jackson Hole and its lakes.

Start of Tour and Stop 1: Flat Creek Valley

Start of tour; 3.7 miles to Stop 2

Begin the tour by driving the Jackson Hole Highway (U.S. 26–89–191) north out of Jackson, Wyoming. The starting point is a parking lot on the east (right) side of the road just after the Wyoming Visitors Information Center and before the highway crosses Flat Creek.

East of the highway is Flat Creek Valley and the National Elk Refuge, established by Congress in 1912 after homesteads and ranches reduced elk wintering grounds in Jackson Hole. Several thousand elk winter on the nearly 25,000-acre refuge.

The big hill west of the highway is East Gros Ventre Butte. (Gros Ventre is pronounced grow-vaunt and rhymes with want.) The butte is about 5 miles long north to south, and only a mile wide. On its other side, out of sight, is a similar hill, West Gros Ventre Butte. Faults run north–south at the east base of each butte, including where you now are standing. During prehistoric quakes, land east of each fault—including Flat Creek Valley—dropped downward relative to land west of each fault, which rose upward to create the buttes. There is no evidence of recent activity on these faults. But ancient quakes made the Flat Creek Valley drop by perhaps hundreds of feet. The stream gradually filled the valley with sediment, creating the flat, lush valley floor that elk find so attractive.

This fault is less than 5 miles long—much smaller than the 40-mile-long Teton fault. But both were created by the same forces: the stretching apart of Earth's crust in the interior western United States, a region known as the Basin and Range Province.

ON THE WAY TO STOP 2

Flowing along a fault: As you proceed north, the highway follows the fault at the east base of East Gros Ventre Butte. Springs that flow out of the ground near the north end of the butte probably exist because of groundwater flowing along the fault. These springs supply water to the fish hatchery, Stop 2.

Turn right to Jackson National Fish Hatchery: *3.2 miles from Stop 1; 0.5 miles to Stop 2.* Proceed to parking lot.

Stop 2: Jackson National Fish Hatchery Parking Lot

3.7 miles from Stop 1; 1.5 miles to Stop 3.

The hatchery is often open for tours.

Below you, the low undulating terrain of the upper Flat Creek Valley was shaped partly by glacial ice that moved south from Yellowstone through Jackson Hole during at least one of three major periods of glaciation in the region during the past 2 million years. The ice once may have been 1,000 feet thick here. East across the valley, look at the west end of the Gros Ventre Range. As you face the fish hatchery, the range's highest peak—11,239-foot Sheep Mountain, also known as Sleeping Indian—is to the northeast on your left. South of it is 10,741-foot Jackson Peak. Before Jackson Hole started dropping downward along the Teton fault 13 million years ago, what now are the Teton and the Gros Ventre ranges formed a continuous high plateau created by the uplift of the Rocky Mountains between 80 and 30 million years ago.

ON THE WAY TO STOP 3

Leave hatchery lot, turn right on main highway: *0.4 miles from Stop 2; 1.1 miles to Stop 3.* The highway now climbs out of the Flat Creek Valley. The cliffs on the west (left) side of the road are andesites, volcanic rocks that probably erupted from a small, nearby volcano 9 million years ago—long before the Yellowstone hotspot arrived beneath Yellowstone National Park and triggered major volcanic eruptions beginning 2 million years ago. The older andesites and several other volcanic features in the area suggest hot rock may have seeped upward and eastward from the hotspot when the hotspot still was beneath Idaho.

The highway proceeds north over grass- and sagebrush-covered landscape known as a glacial outwash plain. Much of the floor of Jackson Hole is glacial outwash—coarse sands and gravels washed out of glaciers by meltwater.

Enter Grand Teton National Park: *0.9 miles from Stop 2; 0.6 miles to Stop 3.* A small sign designates the park boundary.

Stop 3: Grand Teton National Park Sign Turnout

1.5 miles from Stop 2; 14.5 miles to Stop 4.

Stop in the turnout next to the large Grand Teton National Park sign. This sign is 0.6 miles *after* the small park boundary sign.

Look west and northwest from this turnout for our tour's first vista of the spectacular scenery of the 44-mile-long Teton Range. The tallest peak is 13,770-foot Grand Teton, to the northwest. Its summit is some 7,000 feet above the valley floor where you now stand. The mountain to your west is Rendezvous Mountain, and its highest point is 10,927-foot Rendezvous Peak, the southernmost high peak of the Tetons. At its base is Teton Village, a ski resort, and a tramway up the mountain.

The Teton fault is 6 miles west of this turnout. The fault runs about 40 miles north-to-south along the base of the mountains, from the north end of Jackson Lake to south of Teton Village. For as long as 13 million years, a few thousand major earthquakes made the valley drop downward along the fault while the Teton Range rose upward. The oldest rocks in the range are 2.8 billion years old and are found high—but not atop—the central peaks. They were lifted from their birthplace deep underground by both the Teton fault and earlier periods of mountain-building.

As quakes caused Jackson Hole to drop down along the fault, the part of the valley closer to the fault dropped more than the part where you now stand. So the floor of Jackson Hole tilts downward to the west very slightly—less than a degree—from this viewpoint to the base of the Teton Range.

ON THE WAY TO STOP 4

Turn right at Gros Ventre Junction: *1.9 miles from Stop 3; 12.6 miles to Stop 4.* To continue on the main tour, turn right at Gros Ventre Junction onto Gros Ventre Road.

(If you prefer to skip Stop 4, including the Gros Ventre landslide and the east side of Jackson Hole, don't turn right at Gros Ventre Junction. Instead, continue north through the junction and proceed on the main highway [U.S. 26–89–191] for 6 miles to where Antelope Flats Road enters the highway from the right. That intersection is 1 mile past Moose Junction. Pick up the text of this tour guide under the section On the way to Stop 5, Intersection Antelope Flats Road and Jackson Hole Highway.)

If you turned right at Gros Ventre Junction to proceed on the main tour, the bank on your left (north) is the south side of a terrace of gravel and other sediments deposited by the Gros Ventre River, primarily since the glaciers retreated. The top of this terrace is covered by loess, which is dust and other sediments blown off of ancient glaciers by the wind. This terrace extends about 2 miles north. Several gullies cut northeast-to-southwest through the terrace and probably were created by large floods about 5,000 years ago. The river later abandoned its course atop this terrace and shifted to its present position on the southeast (right) side of this road. The river cut down through the old terrace as it shifted, forming the bank that parallels the road on the left.

Gros Ventre campground: *6.4 miles from Stop 3; 8.1 miles to Stop 4.*

Kelly: *8.5 miles from Stop 3; 6 miles to Stop 4.* After driving into Kelly, Gros Ventre Road turns left (north). This village was wiped out by a flood in 1927 when a natural dam ruptured upstream on the Gros Ventre River. The dam was created by the Gros Ventre slide (Stop 4).

Turn right to remain on Gros Ventre Road: *9.9 miles from Stop 3; 4.6 miles to Stop 4.* About 1.4 miles after entering Kelly, turn right to remain on Gros Ventre Canyon Road. Be careful not to miss this turn. Do not go straight.

Kelly Warm Spring and Gros Ventre Canyon: *10.3 miles from Stop 3; 4.2 miles to Stop 4.* About 0.4 miles past the turn, the road passes Kelly Warm Spring, a popular swimming hole with average water temperatures around 80 degrees Fahrenheit. The springs may arise from a buried fault that acts as a conduit for the warm water. After leaving the spring, the road climbs hummocky, aspen-covered hills made of debris deposited by glaciers. You enter Gros Ventre Canyon about 2 miles past the warm springs. Proceed up canyon to Stop 4.

Entering Teton National Forest sign turnout: *12.3 miles from Stop 3; 2.2 miles to Stop 4.* The Gros Ventre landslide is visible up canyon on the mountainside on your right, but keep driving to Stop 4.

Stop 4: The Gros Ventre Slide

14.5 miles from Stop 3; 17.7 miles to Stop 5

Park in the turnout on the right at the Gros Ventre Slide sign. Walk to the viewpoint to see the slide across the canyon. (See Figure 5.9.)

The influence of earthquakes and landslides in the evolution of the Teton region is revealed dramatically at this stop. Here, 50 million cubic yards of rock suddenly gave way on the south side of Gros Ventre Canyon on June 23, 1925, creating one of the world's great landslides. It is bigger than the deadly slide in Madison Canyon, Montana, triggered by the 1959 Hebgen Lake earthquake.

The Gros Ventre Slide, also called the Lower Gros Ventre Slide, probably was triggered by a combination of saturated ground, vulnerable rock, and a series of earthquakes, particularly a moderate jolt the night before the slide. The landslide traveled about 1.5 miles as it dropped 2,100 vertical feet off the canyon's south wall and continued up the north wall near where you now stand. The slide formed a 225-foot-tall natural dam across the Gros Ventre River, creating Lower Slide Lake. The top 50 feet of the dam failed on May 18, 1927, sending a flood downstream. The flood destroyed much of Kelly and damaged ranches and bridges all the way to Wilson. Six people drowned.

ON THE WAY TO STOP 5

Turn around: From the Gros Ventre Slide turnout, turn around and retrace your drive back down Gros Ventre Canyon Road to the T-intersection just past Kelly Warm Spring.

Turn right on unsigned road: *4.7 miles from Stop 4; 13 miles to Stop 5.* From Gros Ventre Canyon Road, turn right onto the unsigned road (left is the continuation of Gros Ventre Road, which is the way you came).

Ditch Creek Fan: You now are driving across one of the largest alluvial fans in the western United States, the Ditch Creek fan. An alluvial fan is formed when a sediment-laden stream pours out of a canyon and spreads out over an open valley, dumping gravel, sand, and other sediment in a fan-shaped deposit that is narrowest and thickest near the canyon mouth and widens and thins as it spreads into the valley. The Ditch Creek fan formed after the glaciers receded 14,000 years ago. It continues to grow today. It was deposited by Ditch Creek, which emerges from the mountains on your right (east). From that point, the fan stretches about 4 miles west until it abuts the northeast side of Blacktail Butte, which is to your left (west). The fan slopes down to the west almost a degree—relatively steep for such a large alluvial fan. Why? Jackson Hole tilts slightly westward due to downward movement along the Teton fault, so drainage from Ditch Creek probably has been accelerated, making the alluvial fan bigger than it would be otherwise.

Turn left on Antelope Flats Road: *7.1 miles from Stop 4; 10.6 miles to Stop 5.* About 2.4 miles past the last turn, you reach Antelope Flats Road. If you turn left or proceed straight, you are on Antelope Flats Road. Turn left. From this point you will have nice views of the Tetons and of Blacktail Butte, only a couple miles ahead and a bit to the left.

Blacktail Butte rises more than 1,000 feet above the surrounding terrain. It basically is a big chunk of bedrock (topped by more recent stream deposits) that was carved by south-moving glaciers. Some of its rocks range from 180 to more than 500 million years old. Elsewhere in Jackson Hole, rocks of the same age are located thousands of feet beneath the valley floor, dropped downward by millions of years of motion on the Teton fault. But here they are exposed well above the valley floor. Why? A possible answer lies due west, at the base of the Teton Range. From there to the north end of the range, the Teton fault runs pretty much north–south, and the fault plane dips east under Jackson Hole. But due west of Blacktail Butte, the Teton fault makes a sudden jog to the southwest, so its buried fault plane dips southeast. Now imagine Jackson Hole dropping downward to the east along one part of the fault, and downward to the southeast along the other. That leaves a wedge-shaped "hole" between the fault planes in the location of Blacktail Butte. This change in fault geometry suggests Blacktail Butte may be an ancestral part of Jackson Hole that failed to drop downward with the rest of the valley. Smaller faults apparently surround the base of Blacktail Butte, supporting the theory. Another theory is that Blacktail Butte is much older, and started as a large block of rock that slid westward off the Gros Ventre Range along a gently dipping fault that formed when the region was being squeezed during the Overthrust period.

Intersection Antelope Flats Road and Jackson Hole Highway (U.S. 26–89–191): *10.4 miles from Stop 4; 7.3 miles to Stop 5.* Turn right onto the main highway. (If you chose to skip Stop 4 earlier in the tour, and instead proceeded north on Jackson Hole Highway, this is where you rejoin the tour, continuing north on the highway.)

Glacier View turnout: *12.5 miles from Stop 4; 5.2 miles to Stop 5.* Active glaciers of the Teton Range are visible.

Teton Point turnout: *14.6 miles from Stop 4; 3.1 miles to Stop 5.* After leaving this turnout, the road passes on top of the western edge of a major Snake River terrace—a nearly flat surface of silt, sand, and gravel deposited by the river before it had cut downward to its present location to your left (west).

Turn left to enter Snake River Overlook: *17.5 miles from Stop 4; 0.2 miles to Stop 5 parking lot.*

Stop 5: Snake River Overlook

17.7 miles from Stop 4; 15 miles to Stop 6

The tree-covered ridge visible to the northwest across the river is a glacial moraine named Burned Ridge. It is a ridge of boulders, gravel, and sand deposited at the toe of a glacier during the most recent stage of glaciation, which ended about 14,000 years ago. Moraine deposits are favorable for vegetation, so many moraines in Jackson Hole are forested. In contrast, the almost flat, brushy, glacial outwash plains have porous soils that cannot hold enough water to support lush vegetation.

One of Jackson Hole's most unusual features can be noted from this overlook: the valley floor's slight westward tilt. The slope is so gentle that it is difficult to observe by looking at the valley floor to the west. But the tilt is evident by looking west and noting that only the tops of tall pine trees are visible on the western edge of the valley. Much of the tilt is due to earthquake movements on the Teton fault, which makes the west side of Jackson Hole drop more than the east side. Perhaps as much as a quarter of the tilt is due to accumulation of sloping gravel and sediment deposits left by streams that once flowed from glaciers.

Across the river to the northeast is a gravel bar named Deadmans Bar after an 1886 incident in which a miner killed three colleagues. He was captured but acquitted on grounds of self-defense.

Farther to your northeast, about 5 miles away, is Signal Mountain, which you will visit during Stop 7. The glacier-sculpted mountain contains west-dipping rock layers. One of those layers, the Huckleberry Ridge tuff, is a brown band of rock wrapping around the southeast side of Signal Mountain. It is volcanic ash deposited by the first gigantic volcanic eruption at Yellowstone 2 million years ago. Like the valley floor, the rocks in Signal Mountain tilt westward because of movement on the Teton fault. Debris dumped by glaciers also is exposed on Signal Mountain.

ON THE WAY TO STOP 6

Exit Snake River Overlook, turn left on Jackson Hole Highway: *0.2 miles from Stop 5; 14.8 miles to Stop 6.*

Spread Creek: *5.9 miles from Stop 5; 9.1 miles to Stop 6.* By the time you reach the bridge over Spread Creek, you are on another alluvial fan. From here, the creek drains into the Snake River, more than a mile away. Measurements reveal the tug of Earth's gravity is slightly weaker than normal here, indicating the rock beneath this area is less dense than normal. This supports the theory that the Spread Creek alluvial fan and earlier deposits of glacial rock filled what once was a deep basin. The basin now is buried under thick sediment, but its prehistoric presence may help explain why the Snake River flows east out of Jackson Lake instead of directly south toward lower ground. With Jackson Lake's southern end dammed by a glacial moraine, it logically would have drained into a basin to the east—a basin later filled with sediments.

Turn left at Moran Junction onto U.S. 89–191–287: *9.6 miles from Stop 5; 5.4 miles to Stop 6.*

Moran Entrance Station: *10.4 miles from Stop 5; 4.6 miles to Stop 6.* Rangers will collect fees for your visit to Grand Teton National Park. It probably will be economical to buy an extended permit because you will leave and re-enter Grand Teton and Yellowstone several times if you follow both tours.

As the highway runs west to Pacific Creek, it passes through hills made largely of debris dumped by glaciers, although some of the higher hills are conglomerate—cemented boulders, rocks, and finer sediments deposited by streams tens of millions of years ago as an ancient highland eroded away.

Oxbow Bend Scenic Overlook: *12.4 miles from Stop 5; 2.6 miles to Stop 6.* A large oxbow-shaped bend in the Snake River formed here because the river has been narrowed a mile or so downstream (behind you) at Pacific Creek. The Snake River once flowed from the south end of Jackson Lake, until that area was dammed by a glacial moraine. That forced the river to flow eastward out of the lake and around Signal Mountain—an unusual flow pattern given the general westward tilt of Jackson Hole (Figure 8.4). This spot provides a classic view of 12,605-foot Mount Moran to the west. Signal Mountain is nearby to the south–southwest.

Turn left at Jackson Lake Junction onto Teton Park Road: *13.5 miles from Stop 5; 1.5 miles to Stop 6.* Turn left at the sign toward Signal Mountain, Jenny Lake,

8.3 ⁓ *The Teton Range and Snake River from the Snake River Overlook. As the river meandered over time, it cut into the valley floor and created stairstep-like terraces. An old terrace is seen as the horizontal line at the base of the embankment stretching across the middle of the photograph.*

8.4 ⁓ *Signal Mountain* (left), *with Mount Moran and other Teton Range peaks* (right). *Signal Mountain tilts to the west because of movements on the Teton fault. The Snake River's Oxbow Bend is in the foreground.*

Moose, and Jackson. Trees on your left are growing on an old landslide that fell from the north side of Signal Mountain after the glaciers retreated.

Stop 6: Jackson Lake Dam

15 miles from Stop 5; 6.8 miles to Stop 7

Stop in the parking lot on the left after you cross the dam.

A log-crib dam first was built here during 1905–1907, raising Jackson Lake 22 feet above its natural level. The early dam failed in 1910. A concrete dam was built in stages between 1911 and 1916, raising the maximum lake level another 17 feet, to 39 feet above the lake's natural elevation. The dam impounds irrigation water for agriculture downstream the Snake River in Idaho and also provides flood control. In 1976, on the west side of the Teton Range, the Teton River's Teton Dam ruptured, causing a catastrophic flood that killed eleven people and caused $900 million in damage. The disaster prompted the U.S. Bureau of Reclamation to study other dams. It found Jackson Lake Dam was susceptible to failure during a nearby earthquake larger than magnitude 5.5. The dam was upgraded during 1986–1989, and the Bureau of Reclamation now believes it can withstand the "maximum credible earthquake"—a magnitude-7.5 quake on the Teton fault.

Ancestral Jackson Lake formed in a trough at the base of the Teton Range, probably soon after the Teton fault became active about 13 million years ago. The trough was created by the downward movement of Jackson Hole along the Teton fault, which made the ground sink more at the base of the range than on the east side of the valley. The lake now includes a deeper western trough, west of here but out of view, and a shallower, shorter eastern trough, which begins at the dam site and stretches 3 miles to the southwest. The western trough, initially created by faulting, was scooped out and deepened by glaciers that flowed south from the Yellowstone Plateau during the most recent glacial advance, which ended roughly 14,000 years ago. The lake's eastern trough, which you see from the dam, was carved by an earlier glacier flowing down Pacific Creek's valley. Water is 437 feet deep in Jackson Lake's western trough and 142 feet deep in the shallower eastern trough. Both troughs were deeper at one time, but have filled with sediments.

The muddy sediments in Jackson and Yellowstone lakes are unusual because they contain the silica-rich remains of microscopic, single-celled algae called diatoms. When they were alive, the diatoms obtained nutrients from the silica-rich rhyolite lavas that erupted at Yellowstone.

Some researchers have hypothesized the Teton fault reaches the Earth's surface beneath the waters of Jackson Lake. But pictures of lake-bottom sediments made with sound waves show muds and deeper sediments beneath Jackson Lake have not been broken or folded by faulting. Instead, the fault intersects the ground west of the lake at the base of the Teton Range. There, the fault has created scarps—or steep, cliff-like slopes—where the mountains rose up and Jackson Hole dropped down during the most recent prehistoric major quakes.

ON THE WAY TO STOP 7

Continue south along Jackson Lake. The road passes through thick stands of pines growing on glacial moraines.

Signal Mountain Lodge: *1.8 miles from Stop 6; 5 miles to Stop 7.* A possible lunch spot halfway through the tour.

Turn left on Signal Mountain Road: *2.8 miles from Stop 6; 4 miles to Stop 7.* From this turn, a winding road climbs up Signal Mountain, which rises almost 800 feet above the floor of Jackson Hole. Stop 7 is *not* the summit parking lot. Stop 7 is the Jackson Point Overlook 4 miles up the mountain from this turn and about a mile before the summit. The lot is just past a hairpin curve. Watch out for bicyclists and oncoming traffic on this narrow road.

Stop 7: Jackson Point Overlook on Signal Mountain

6.8 miles from Stop 6; 6.2 miles to Stop 8

Once at the Jackson Point parking lot, walk a short trail to the overlook.

Signal Mountain is the most important stop on this tour because of its panoramic views of the Teton Range, Jackson Lake, Jackson Hole's glaciated landscape, the Snake River drainage, and the Pitchstone Plateau, which is the southernmost part of the Yellowstone Plateau.

To the south is much of the valley of Jackson Hole. East Gros Ventre Butte also is visible.

Turn clockwise a bit. To the southwest is the relatively gentle southern Teton Range.

To the west–southwest and west is the highest, most jagged part of the Teton Range, rising to 13,770-foot Grand Teton. Other tall peaks visible from here include Mount St. John and Mount Moran (Figure 8.5).

To the northwest, the north end of Teton Range becomes smoother and lower, and is capped by the Huckleberry Ridge tuff, a rock layer formed from the thick layer of volcanic ash and debris spewed skyward by the eruption of the Yellowstone caldera 2 million years ago.

Observing the Teton Range from Signal Mountain provides a great view of the full extent of the Teton fault's power to change the landscape by lifting up a 44-mile long mountain range and making Jackson Hole drop downward. And the present 7,000-foot difference in elevation from the valley floor to the top of Grand Teton represents only part of the fault's power. During the past 13 million years, the mountains rose and the valley dropped by about 23,000 feet. But sediments filled the valley as it sank, so only a third of this movement is now obvious as one looks from the valley floor to the mountain tops.

Notice how the highest peaks of the Tetons tower above and only a few miles west of Jackson Hole. Yet canyons rising from Jackson Hole extend much farther west into the Tetons. In other words, the range's east–west drainage divide is located about 2.5

8.5 ~ The Teton Range and Jackson Hole from the Jackson Point Overlook on Signal Mountain.

miles west of the highest peaks. So snow and rain falling onto the high peaks drain eastward into Jackson Hole. This pattern is unusual. In most ranges, the tallest peaks define the drainage divide. But the east side of the Teton Range rose rapidly along the Teton fault while the west side remained relatively fixed. So the east side of the range eroded more quickly than the west side. As a result, canyons cut westward through the area of the highest peaks, creating a drainage divide that does not coincide with the highest portions of the range.

Now look north. The long, flat, tree-covered ridge in the distance is the Pitchstone Plateau, which attains elevations of nearly 9,000 feet and constitutes the southern portion of the Yellowstone Plateau. The Pitchstone Plateau was formed 70,000 years ago by the youngest of about thirty lava flows that erupted after the last cataclysmic eruption of the Yellowstone caldera 630,000 years ago.

On the far northeast skyline is the Absaroka Range, which was formed by volcanic eruptions that began 50 million years ago when the area was near the ancient west coast of North America.

East and southeast of Signal Mountain are the Mount Leidy Highlands and the high gray peak of 10,326-foot Mount Leidy. They are composed of ancient seafloor sands and shales deposited in the area between roughly 100 and 70 million years ago, then uplifted by the subsequent squeezing of Earth's crust that built the Rockies. To the south and east of Mount Leidy is the Gros Ventre Range with its highest peaks on the southeast skyline, including Sheep Mountain and Jackson Peak. The core of the Gros Ventre Range is 2.5 billion years old.

Now look to the south again to examine the effects of glaciers on Jackson Hole. Imagine 2,000-feet-thick glaciers slowly flowing through this valley, thinning, and then tapering out beyond the present site of Jackson, Wyoming. The overlook where you stand would be under more than 1,000 feet of ice. During the past 2 million years, glaciers moving through Jackson Hole buried Signal Mountain several times. During the last glacial stage, which ended about 14,000 years ago, three huge lobes of ice entered the valley from the Snake River on the north, Pacific Creek on the northeast, and the Buffalo Fork of the Snake on the east. These glaciers originated from a vast ice sheet that covered Yellowstone and the Absaroka Range.

Burned Ridge is the pine-covered ridge 5 miles southwest of this overlook. It extends from northwest to southeast—perpendicular to the giant lobe of ice formed by glaciers that once emerged from Buffalo Fork and Pacific Creek. Burned Ridge is a terminal moraine, or pile of debris dumped at the toe of the glacial lobe. Another terminal moraine deposited by a more recent glacier dammed the southern end of Jackson Lake.

The bumpy plains southeast, south, and southwest of Signal Mountain are known as knob-and-kettle topography—small hills of glacial debris, plus ponds or potholes left when the glaciers receded (Figure 8.6). Huge blocks of underground ice melted, leaving the potholes or kettles.

For a time after the big glaciers eventually retreated from Jackson Hole, smaller alpine glaciers continued to flow off the Teton Range, helping to carve the classic U-shaped canyons that descend into the valley. Moraines and hummocky topography at the toes of these glaciers dammed Jenny Lake and shaped the conifer-covered hills at the base of the mountains. About a dozen small glaciers still persist high in the Tetons.

ON THE WAY TO STOP 8

Turn around: From the Jackson Point Overlook, turn around and drive 4 miles back down Signal Mountain Road.

Turn left on Teton Park Road: *4 miles from Stop 7; 2.2 miles to Stop 8.* The highway soon crosses open sagebrush-covered terrain composed of gravel and other

8.6 ~ *Glacial outwash plain of northern Jackson Hole seen from the Jackson Point Overlook on Signal Mountain, with 10,326-foot Mount Leidy in the background. Light areas in the foreground show hummocky "potholes" terrain left by glaciers. The potholes can fill with water like Cow Lake* (center).

glacial outwash debris deposited some 14,000 years ago by the most recent glaciers in Jackson Hole. Stairstep-like terraces here were formed when ancient streams once flowed south from Jackson Lake.

Potholes Turnout: *5.6 miles from Stop 7; 0.6 miles to Stop 8.* Views of hummocky "potholes" terrain.

Stop 8: Mount Moran Scenic Turnout

6.2 miles from Stop 7; 4.5 miles to Stop 9

Mount Moran is the northernmost of the high peaks in the Teton Range at 12,605 feet and has a flat, eroded top. The peak is made of 2.8-billion-year-old rock called gneiss and 2.4-billion-year-old granites. Long before the modern Teton Range was lifted upward, this rock was deep underground. About 1.5 billion years ago, molten rock squeezed upward into the granites, then solidified into a rock named diabase, forming a dark, vertical layer or "dike" visible in Mount Moran (Figure 8.7).

The peak is topped by a thin layer of another kind of rock, named the Flathead sandstone, formed when the area was once a seafloor. It is about 540 million years old. The same layer is buried deep beneath Jackson Hole. The vertical distance between the Flathead sandstone atop Mount Moran and beneath Jackson Hole is roughly

8.7 ∾ *Mount Moran with a vertical black dike of iron-rich rock. The dike formed 1.5 billion years ago when molten rock was injected into older rock that now forms Mount Moran. All the rocks were about 6 miles underground at the time, long before the Teton Range was uplifted. Falling Ice* (center) *and Skillet* (right) *glaciers help shape the mountains.*

30,000 feet. About 23,000 feet of that vertical movement happened during the past 13 million years as the mountains rose and the valley dropped along the Teton fault. The remaining difference was caused by earlier episodes of mountain-building in the ancestral Teton Range.

Also visible from the spot is Falling Ice Glacier, one of the remaining active glaciers in the Tetons.

ON THE WAY TO STOP 9

Pines on glacial debris: *0.6 miles from Stop 8; 3.9 miles to Stop 9.* The roadway passes into stands of pine growing on Burned Ridge, which is a terminal moraine of debris dumped at the toe of a glacier.

Ancient river channel: *2.6 miles from Stop 8; 1.9 miles to Stop 9.* After leaving the forest, Teton Park Road crosses an ancient Snake River channel that once drained the southwest end of Jackson Lake before a glacial moraine blocked that outlet. The road climbs a river terrace on the west side of the channel.

Mountain View turnout: *3.3 miles from Stop 8; 1.2 miles to Stop 9.*

Turn right at North Jenny Lake Junction: *3.7 miles from Stop 8; 0.8 miles to Stop 9.* The road crosses outwash debris deposited by streams flowing from a glacier. Vegetation now covers the outwash.

Stop 9: Cathedral Group Scenic Turnout

4.5 miles from Stop 8; 4.3 miles to Stop 10

One of the most dramatic young fault scarps in the western United States is visible to the west of this viewpoint. A scarp is a steep bank or slope—often almost a cliff—formed by movement on a fault. In this case, Jackson Hole dropped downward along the Teton fault while the Teton Range moved upward. The steep scarp visible from here was formed by earthquakes only since the glaciers receded about 14,000 years ago, but it illustrates how quakes in most recent prehistoric time helped build the Teton Range during the past 13 million years.

The scarp is located just west of String Lake. It measures 124 feet from top to bottom. But the scarp was created by actual vertical movements on the Teton fault of only about 75 feet. That represents as many as a half dozen magnitude-7.5 quakes or perhaps twenty magnitude-7.3 quakes since the glaciers receded about 14,000 years ago. Yet only two of those quakes happened in the past 8,000 years, suggesting the fault may be long overdue for another disaster.

Why is the scarp taller than the actual vertical movements during quakes along the Teton fault? During or after a big quake, large wedges of ground can slump downhill from the base of the scarp, adding to its height. Such slumping is complicated, and is related to the westward tilt of Jackson Hole's valley floor.

The tilt is apparent from this stop. As you look west, you see only the tops of trees near String Lake. The trees and the lake sit in a trough created largely by earthquakes. During major quakes, the part of the valley closest to the fault and the mountains drops downward more than land farther east (behind you). As a result, the entire valley floor tilts westward. In addition, smaller faults form in the valley floor a few hundred yards east of the Teton fault. These faults dip westward and intersect the east-dipping Teton fault. That means the part of the valley floor nearest the Teton fault literally breaks off to form a "wedge" of ground separate from the rest of the valley floor. This wedge moves downward and rolls a bit westward, amplifying both the height of the fault scarp and the depth of the trough at the base of the mountains. Along with lake-damming glacial moraines, this trough—which is about 95 feet deeper than the glacial outwash plain behind you—explains why Leigh, String, Jenny, and other lakes formed along the fault, and why streams connecting the lakes drain south, not east, away from the Teton Range.

ON THE WAY TO STOP 10

Overlook on east shore of Jenny Lake: *2.2 miles past Stop 9; 2.1 miles to Stop 10.* Continue on the road that runs over glacial moraine and other debris and skirts the east side of Jenny Lake. The overlook is a parking lot on the right. The down-dropped trough in which Jenny Lake formed is evident from the overlook, which also provides a view up Cascade Canyon.

Turn right onto Teton Park Road at South Jenny Lake Junction: *3.2 miles from Stop 9; 1.1 miles to Stop 10.*

Turn right into South Jenny Lake Area: *3.8 miles from Stop 9; 0.5 miles to Stop 10.* Then proceed to the southwest end of the parking lot, which is Stop 10.

Stop 10: South Jenny Lake Area, Cascade Canyon View, and Optional Hike to Glacial Moraine and Teton Fault

4.3 miles from Stop 9; 3.3 miles to Stop 11

This stop includes a visitor center, rest rooms, grocery store, and other amenities, making it a good place for a rest break. Even though it has been paved, you are parked atop a moraine of glacial debris that was dumped into Jackson Hole by a glacier that once flowed east out of Cascade Canyon.

The mouth of Cascade Canyon is visible from here. It is a classic U-shaped canyon, indicating it was carved by one of the most recent glaciers to flow east from the Teton Range. Streams carve V-shaped canyons. To the west, just south of Cascade Canyon and high above you, is 12,325-foot Teewinot Mountain.

Boat rides are available across the lake to Inspiration Point. But to walk across the south Jenny Lake moraine to the Teton fault, consider an optional 2.6-mile round-trip hike (or 4.4 miles if you go all the way to Inspiration Point). Start either at the trail that passes the boat dock or the trail at the south end of the parking lot. Both trails cross Cottonwood Creek, then merge and lead around the south end of Jenny Lake. Ignore the trails to Moose Pond. After rounding the lake's south end, the trail turns northwest and begins to climb the moraine, a ridge of debris deposited by the glacier that once flowed here. At a junction where the right fork follows the lake shore, take the left fork onto the horse trail instead. This trail climbs for a few hundred yards, then drops, then makes a switchback where it begins to climb a scarp formed in the moraine by the Teton fault. The small valley between the trail drop and the switchback is called a graben. As the Teton Range rose during big quakes, a "wedge" of the valley floor right next to the fault slumped downward along both the east-dipping Teton fault and a smaller west-dipping fault, creating the trough-like graben.

Turn around here for the 2.6-mile round-trip. Or continue north less than a mile to Hidden Falls. Then proceed a short distance northeast to Inspiration Point for a spectacular view of Jenny Lake, the other end of the moraine on the east side of the lake, and the glacial outwash plain farther east. You also can look west into Cascade Canyon and active glaciers on the north side of Mount Owen. Return the way you came or by the lower trail on the lake shore.

ON THE WAY TO STOP 11

Leave the parking lot, turn right on Teton Park Road: *0.4 miles past Stop 10; 2.9 miles to Stop 11.* Drive out of the South Jenny Lake Area parking lot and turn right onto Teton Park Road. The road traverses a flat plain created by sediments that washed out of glaciers. To your left (east), is Timbered Island, a tree-covered moraine of debris dumped along one edge of a south-flowing glacier. To your right (west) are other moraines formed by glaciers that once flowed from the Teton Range. You will pass at least two unsigned turnouts before Stop 11.

Stop 11: Teton Glacier Turnout

3.3 miles from Stop 10; 4.0 miles to end of tour

This stop provides views of 13,770-foot Grand Teton and other high peaks. Grand Teton's rugged topography was formed by the uplift, erosion, and intense glaciation of 2.4-billion-year-old rock. A dike, or vertical layer of rock, is visible extending downward from the top of Middle Teton. It is one of several dikes that formed when molten rock squeezed upward about 1.5 billion years ago into the older rocks that now form the Teton Range.

The Teton Glacier and the terminal moraine at the end of the glacier also can be seen to the right of Grand Teton. This glacier has been receding during most of historic time, as have others in the Teton Range.

Directly west of this stop, you can see the tops of trees along Cottonwood Creek. Like String, Jenny, and other nearby lakes, the stream sits within the trough along the Teton fault.

ON THE WAY TO THE END OF THE TOUR

Windy Point Turnout: *2.7 miles from Stop 11; 1.3 miles to the end of the tour.* The tree-covered bench west of the road is Windy Point, the east edge of a moraine dumped by a glacier that once flowed out of the Teton Range. Windy Point is topped by a layer of loess, or windblown glacial soil. The visible bank or slope formed when the Snake River once flowed here and cut through the moraine. Windy Point now diverts Cottonwood Creek from its southerly course and forces it to flow southeast into the Snake. After leaving Windy Point, the road continues south and drops into young sediments deposited by the modern Snake River.

End of Grand Teton Tour, Moose Entrance Station

4.0 miles from Stop 11. End of tour

This tour ends as you exit Grand Teton National Park's Moose Entrance Station. You have four options from here:

1. About a third of a mile past the entrance station, turn left to see the Moose Visitor's Center and park headquarters. Maps and books about the park are available.
2. Also a third of a mile past the entrance station, you may turn right onto the Moose–Wilson road for an almost 9-mile drive to Teton Village and the optional aerial tramway ride up Rendezvous Mountain. (See the following description.) The tramway usually closes in the late afternoon. A 1.3-mile stretch of the Moose–Wilson Road is unpaved gravel.
3. One mile past the entrance station, reach Moose Junction and turn right on Jackson Hole Highway (U.S. 26–89–191) and drive about 13 miles to accommodations in Jackson, Wyoming.

4. One mile past the entrance station, turn left on Jackson Hole Highway and proceed 18 miles northwest to Moran Junction, then left on U.S. 89–191–287 through the Moran Entrance Station and north to Colter Bay Village or other accommodations in northern Grand Teton National Park. That will put you closer to the start of the Yellowstone tour.

Optional Trip: Aerial Tramway at Teton Village

An excellent way to get an overview of southern Jackson Hole and the south end of the Teton Range is to ride the aerial tramway from the Jackson Hole Mountain Resort at Teton Village, Wyoming.

Teton Village may be reached either by driving 4 miles east from Jackson on Wyoming 22, crossing the Snake River, and then turning right (north) on Moose–Wilson Road and proceeding several miles, or by driving from near the Moose Entrance Station and heading south on the Moose–Wilson Road.

Dates vary, but the tramway operates from late spring through early fall, then again during winter ski season; it is closed much of the spring and late fall. Winter operating hours depend on weather. During late spring, the tram closes in the late afternoon, but operating hours extend into early evening during summer.

The tram climbs from Teton Village, at roughly 6,300 feet, to a 10,450-foot station on Rendezvous Mountain, an elongated mountain topped by 10,927-foot Rendezvous Peak, located a few miles southeast of the upper tram station.

As the tram begins the ascent from Teton Village, it crosses the Teton fault, which cuts across a meadow about 100 yards up the slope and west of the tram building. Signs of the fault have been obliterated by development at the resort.

Although rocks near the bottom end of the tramway are covered by vegetation, the tram first crosses young alluvial fan deposits carried out of the mountains by streams. Next, the tram crosses 2.8-billion-year-old Precambrian rock called gneiss, then across progressively younger sea-bottom deposits of sandstone, shale, and, at the summit, limestone. The mountain-building power of the Teton fault and earlier episodes of uplift are evident. The limestone, now almost 2 miles above sea level, was deposited on a seafloor more than 360 million years ago.

The upper tram station provides a stunning view of the south end of the Teton Range. Imagine how large earthquakes during millions of years made the mountains rise up along the Teton fault while Jackson Hole dropped downward.

In the Teton Range north of here, the top of the range and layers of rock within the mountains are tilted gently westward. Imagine a hinge at the west base of the Tetons. As quakes along the range's east base made the mountains rise there, the west base stayed pretty much in place, tilting the range and its rocks westward.

Look northeast and you will see how the hills and Gros Ventre Range east of Jackson Hole also dip westward, diving beneath the sediment-filled valley floor. The dipping rock layers east of Jackson Hole once were connected with the same layers in the Teton Range. But the fault broke the layers and lifted the Teton Range upward. So rock layers that now are deep beneath Jackson Hole were broken, with their corresponding layers now found thousands of feet higher in the Teton Range.

The southern extent of glacier-carved landscape in the region can be seen by peering down into Jackson Hole. During the past 300,000 years—and perhaps 2 million years—glaciers repeatedly flowed into the valley, both from the Teton Range and, more important, from the Yellowstone region to the north. These glaciers at times were as much as 2,000 feet deep, perhaps more, although they were that thick farther north of here and perhaps 1,000 feet thick here. From the top of the tram to the valley floor is more than 4,000 feet. Imagine the valley below you filled with ice a quarter of the way up to where you stand.

Also visible from the top of the tramway is the Snake River and its flood plain. People have built levees to contain the river. Homes dot the Snake River Valley at the base of the Teton Range near here. Some of them sit on the west side of the valley, part of which is below the level of the Snake River. The river has built up its own bed by dumping sediments, and because prehistoric quakes caused ground close to the Teton fault to drop downward more than the land farther east. A major quake today could damage levees, allowing the river to flow downhill to the west, inundating low-lying ranching and residential areas.

Yellowstone Tour 9

How to Use This Tour Guide

This tour of Yellowstone National Park and the Hebgen Lake earthquake area begins at Yellowstone's south entrance, just north of Grand Teton National Park. Those entering Yellowstone from other directions may start this tour at any stop that is convenient. For that reason, we have not shown cumulative mileage for this trip from start to finish. Instead, we provide cumulative mileage only from one stop to the next, and for points of interest between them. (Figure 9.1. See also Figure 1.4 for a view of the region's topography.)

Two long days or three less hectic days are required for this 251-mile tour. For those with limited time, some time-saving options are included, such as skipping West Yellowstone, Montana, and the Hebgen Lake earthquake area (Stops 7–12) or omitting Mammoth Hot Springs and other stops on the northern end of the park (Stops 14–16).

If you want to spend only two days in Yellowstone and its environs, start early the first day (136 miles) and continue to Stop 12, West Yellowstone, Montana, where overnight accommodations are available. During the second day, visit Stops 13–20 (115 miles).

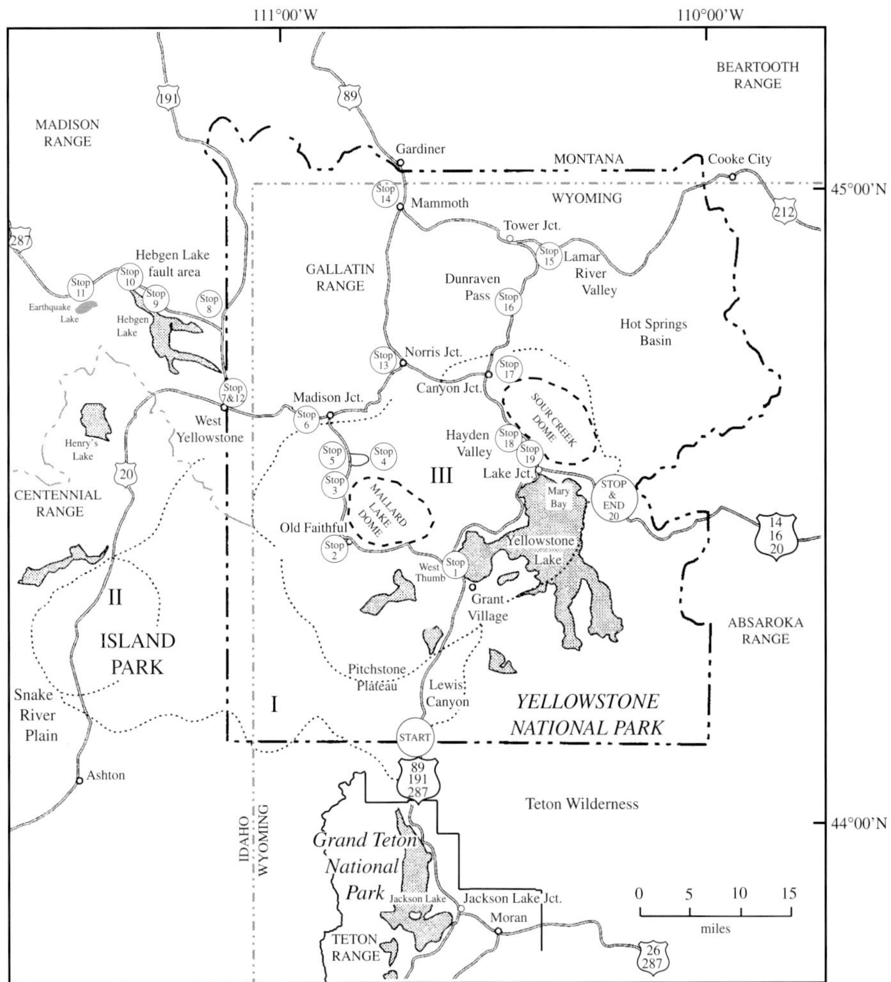

9.1 ~ *Index map with numbers for each stop during this chapter's Yellowstone National Park tour, which begins at the park's south entrance and ends at Lake Butte above Yellowstone Lake. Dotted lines show outlines of ancient eruption calderas.*

If you wish to divide this chapter's tour into three days, begin at Stop 1 and proceed through Stop 7, also West Yellowstone (79 miles). Spend the night there. Tour the Hebgen Lake area, Stops 8–11, the second day and return to West Yellowstone, Stop 12, the second night (57 miles). Cover Stops 13–20 the third day (115 miles).

The tour ends inside Yellowstone, so you may want to spend the night in the area if you have a long drive home.

The driving tour of Yellowstone is passable only from late spring to fall, when park roads are open. West Yellowstone, Montana, and the Hebgen Lake area, Stops 7–12, are accessible year-round, although winter snow covers most geological features.

Vehicle odometers vary, sometimes significantly, so mileages should be taken as approximate. Some visitors may choose to drive part or all of these tours in a direction opposite to the one we use. For that reason, we also provide reverse mileage between each stop and the sights between stops.

The stop-to-stop mileage in this tour includes distances from one parking location to the next, not simply to the entrances of parking lots.

For background before driving this Yellowstone Tour, we suggest reading Chapter 3, Cataclysm! and Chapter 4, How Yellowstone Works.

Introduction

Two million years of violent volcanic eruptions—fueled by the subterranean presence of hot and molten rock in the Yellowstone hotspot—produced the spectacular features seen today at Yellowstone National Park. The park contains geyser and hot-springs systems of extraordinary size and expanse. These attractions are spread throughout a thickly forested, mountain plateau 8,000 feet above sea level.

Visitors often are surprised by the Yellowstone Plateau's relatively flat, undulating topography and its paucity of high mountains. The plateau exists, in part, because the entire region bulged upward over the past few million years as this part of North America moved over the Yellowstone hotspot, a rising plume of hot and molten rock within Earth's mantle and crust.

During the past 2 million years, the hotspot generated three incredibly large eruptions that blew huge craters in the ground: twice at Yellowstone and once farther southwest at Island Park, Idaho. These giant volcanic craters, known as calderas, erupted on a scale unknown in modern time. Huge amounts of molten rock exploded skyward, destroying mountain ranges, incinerating anything within dozens of miles, and dumping volcanic ash over half of North America. Much of central Yellowstone National Park—including much of Yellowstone Lake—sits within the 45-by-30-mile caldera created by the last big blast about 630,000 years ago.

The rim of this vast crater is apparent in some locations but not in many other places where the caldera was filled in by smaller lava and ash flows that erupted since the entire caldera last exploded. Glacial ice about 3,500 feet thick covered Yellowstone repeatedly during the Ice Age. The glaciers scoured and smoothed the landscape, putting finishing touches on the modern Yellowstone Plateau, including Yellowstone Lake.

Yellowstone remains alive as a tremendous amount of heat flows from the ground. The entire floor of the caldera huffs upward and puffs downward over decades as molten rock, hot water, and steam move through conduits at various levels beneath the caldera, from a few miles deep in the crust to perhaps a couple hundred miles deep in the mantle. The shallowest hot water and steam erupt to the surface as ever-changing geysers, springs, mud pots, and steam vents. Movements of molten rock and superheated water at somewhat greater depths can generate thousands of earthquakes, mostly small to moderate ones within the caldera. Other quakes, ranging from small to disastrous, occur because Earth's crust is being stretched apart in a wide area of the western United States. Yellowstone sits on the eastern edge of that zone, known as the Basin and Range Province.

Effects of the Hebgen Lake earthquake of 1959 are the focus of the middle part of this tour. It was a major quake at magnitude 7.5 that set off a gigantic landslide and killed twenty-eight people. Underground movement of molten rock as well as the stretching of Earth's crust may have played a role in causing the disaster.

This tour also will provide views of spectacular geologic features such as the Grand Canyon of the Yellowstone, expansive Yellowstone Lake (one of the highest large lakes in the world), and steaming geyser basins.

Start of Tour: South Entrance, Yellowstone National Park

Start of tour; 21.6 miles to Stop 1

This tour starts at the south entrance to Yellowstone National Park, accessible by driving north from Jackson Hole on U.S. 89–191–287. It is located approximately 46 miles north of Moose, the ending point for the Grand Teton Tour in Chapter 8, or 22 miles north of Jackson Lake Junction.

ON THE WAY TO STOP 1

Start ascending Yellowstone Plateau: *1.6 miles from start of tour; 20 miles to Stop 1.* The highway starts to climb a 5-mile-long grade. You literally are driving up the southern flank of a huge volcanic crater known as the Yellowstone caldera, a 45-by-30-mile hole in the ground that last exploded about 630,000 years ago. You also are, in effect, climbing onto the bulge in Earth's surface created by hot and molten rock in the underlying Yellowstone hotspot.

Lewis Canyon: *4.2 miles from start of tour; 17.4 miles to Stop 1.* Lewis Canyon comes into view. If you want to stop, a large turnout is 2.4 miles past this point. Lewis Canyon cuts about 350 feet deep into the rock formation visible on the east side of the canyon. It is solidified volcanic rock. The downward movement of land south of here, namely, the sinking of the Flagg Ranch area along the Teton fault, may have accelerated the speed and slope of the Lewis River and thus the cutting of Lewis Canyon. Once you have been driving along the canyon for a few miles, the Pitchstone Plateau lava flow is along the west side of the road. The light colored lava can be seen in road cuts. The lava flowed during an eruption about 70,000 years ago—the most recent such eruption in Yellowstone. Since the caldera blew up about 630,000 years ago, it was gradually filled in by a series of these "post-caldera" eruptions. One-third of Yellowstone—including the Lewis Canyon area—was devastated by forest fires in 1988. Yet the energy released by those fires was minuscule compared with even the smallest of Yellowstone's volcanic eruptions.

Lewis Falls optional stop: *9.9 miles from start of tour; 11.7 miles to Stop 1.* Scenic opportunity.

Lewis Lake: *11.2 miles from start of tour; 10.4 miles to Stop 1.* Lewis Lake formed in a depression created when glaciers flowed south over this area and scoured the landscape.

Continental divide and south caldera boundary: *17.3 miles from start of tour; 4.3 miles to Stop 1.* As you pass the continental divide at 7,988 feet above sea level, you also are crossing the rim of the Yellowstone caldera and entering the caldera.

Turn east (right) at West Thumb Junction: *21.2 miles from start of tour; 0.4 miles to Stop 1.*

Turn right again into West Thumb Geyser Basin parking lot: *21.3 miles from start of tour; 0.3 miles to Stop 1.*

Stop 1: West Thumb Geyser Basin and Yellowstone Lake

21.6 miles from start of tour; 19.7 miles to Stop 2

Walk the boardwalk through the West Thumb Geyser Basin.

Like many geyser basins in Yellowstone, West Thumb geysers exist because of volcanic faulting. The basin is on the inside edge of the 45-by-30-mile Yellowstone caldera. When the caldera last erupted 630,000 years ago, the floor of the giant crater dropped

9.2 ~ *Fishing Cone, on the shore of Lake Yellowstone's West Thumb, was produced by minerals deposited by water flowing from the hot spring in the cone.*

downward more than 1,000 feet along a ring-shaped fault. The heat that drives West Thumb's geysers and hot springs is believed to rise from hot rocks in the crust via fractures created by the caldera eruption and subsequent smaller, explosive eruptions.

Offshore is Yellowstone Lake's West Thumb, a circular basin connected to the rest of the lake by a channel. The lake's West Thumb occupies a crater formed by an eruption about 150,000 years ago. The area around West Thumb also has experienced several lesser but still impressive eruptions called phreatic or steam eruptions. Water flowing underground through hot rock is converted to steam. If there is no outlet for this steam to reach the surface, pressure builds until a steam explosion occurs, blowing out overlying rock to create a crater.

Measurements show hot rock still exists beneath West Thumb. The amount of heat flowing upward from the lake bottom is about thirty-five times greater than the heat flow from the ground elsewhere in the Rocky Mountains. A walk through West

Thumb Geyser Basin provides abundant evidence: Hot springs, small dormant geysers, and mud pots named Thumb Paint Pots. Like other mud pots in Yellowstone, they were produced when rock was dissolved by acids in hot water and steam.

Scientists have noted the temperatures and water discharges from West Thumb's geysers and hot springs vary more than other hydrothermal features in Yellowstone, possibly because of periodic infiltration of water from the lake and earthquakes that shake the basin's plumbing system.

West Thumb's hot springs, like others in Yellowstone, are brightly colored by cyanobacteria—commonly called blue-green algae—and other microbes that thrive in hot water. Different microbes grow at various temperatures, so a pool's color can be used to estimate its temperature.

On the shore of Yellowstone Lake is Fishing Cone, where early park explorers supposedly caught fish in the lake, then cooked them in the hot water flowing from the hot spring (Figure 9.2). Such cones are made of minerals deposited by hot springs and geysers. Fluctuating lake levels often submerge Fishing Cone.

ON THE WAY TO STOP 2

Turn west (left) as you leave the West Thumb Geyser Basin parking lot: *0.1 miles from Stop 1; 19.6 miles to Stop 2.*

Turn north (right) at West Thumb Junction: *0.3 miles from Stop 1; 19.4 miles to Stop 2.* Proceed toward Old Faithful.

Ascending the Dry Creek flow: *0.9 miles from Stop 1; 18.8 miles to Stop 2.* The road continues west and climbs 700 feet onto the Dry Creek lava flow, which erupted about 162,000 years ago. The highway crosses the continental divide twice before reaching the Upper Geyser Basin. As you drive to Old Faithful, the road cuts through lava flows in a couple of locations.

Turn right to enter Old Faithful area: *17.8 miles from Stop 1; 1.9 miles to Stop 2.* Follow the road to the parking lot next to the Old Faithful Visitor Center and Old Faithful Lodge.

Stop 2: Old Faithful and Upper Geyser Basin

19.7 miles from Stop 1; 7.2 miles to Stop 3

Plan on spending at least 2 hours at this stop to see dozens of hot springs and geysers, including one of Old Faithful's eruptions. Tour the Old Faithful Visitor Center, con-

9.3 ~ Beehive Geyser in Upper Geyser Basin. (Rick Hutchinson)

sult the list of geyser eruption times predicted by the National Park Service, and buy one of The Yellowstone Association's inexpensive Old Faithful brochures, which includes a trail map and descriptions of individual geysers and springs.

See Old Faithful, then take the Geyser Trail to begin a roughly 3-mile loop walk on the boardwalk across Geyser Hill, then past Giantess, Grand, Giant, Grotto, and Riverside geysers to Morning Glory Pool. Then return to Old Faithful by way of Daisy and Castle geysers. Shorter loops are available. Stay on the boardwalk to avoid damaging vegetation and brittle, delicate mineral deposits from the hot springs and geysers.

Geyser Hill purportedly contains the one of the world's greatest concentrations of geysers. Giantess usually erupts only a few times a year, rising to heights greater than Old Faithful. Giantess' underground water supply is connected to other geysers on the hill. When it erupts, the other geysers erupt less frequently and less vigorously. Vault Geyser was dormant since 1988 until it was reactivated by a 1998 quake.

Old Faithful and Upper Geyser Basin are located in a flat area drained by the Firehole River. This flat area lies between two lava flows that didn't quite converge. The 115,000-year-old Summit Lake lava flow is the forested slope across the highway to the south. It helps form Yellowstone's Madison Plateau. The 151,000-year-old Mallard Lake flow is the timber-covered hill to the north. The Mallard Lake flow is more than just a lava flow. The lava sits atop one of two upward-bulging "resurgent domes" on the floor of the Yellowstone caldera. Molten rock or magma beneath the caldera pushed the caldera floor upward under this area for much of the twentieth century, making the hill formed by the lava flow bulge upward more than it would have otherwise. The Mallard Lake dome is about 7 miles long and 5 miles wide.

The geysers and hot springs in Upper Geyser Basin reach the surface because the land here was not covered by the lava flows. Water seeps into the ground from snow that blankets Yellowstone each winter and then melts. Porous lava flows surrounding Upper Geyser Basin allow the water to percolate downward until it becomes superheated and eventually erupts through geysers or hot springs.

Geysers are eruptions of hot steam and water expelled from shallow chambers. When the pressure of the hot, underground water becomes too great, water and steam erupt to the surface through narrow "pipes" in the geyser basin's subterranean plumbing system. It is similar to the pressure buildup within a pressure cooker. Geysers can erupt dozens to several hundred feet skyward. The hot water beneath Upper Geyser Basin reaches temperatures of more than 400 degrees Fahrenheit—well above boiling. The water cools to about 200 degrees Fahrenheit as it spurts from the geysers.

The time interval between eruptions of any geyser varies. So do the height and volume of water spewed from a geyser. Even Old Faithful isn't what its name implies. It once erupted roughly hourly, but the intervals between outbursts became longer after the 1959 Hebgen Lake, Montana, earthquake northwest of Yellowstone and again after the 1983 Borah Peak earthquake in Idaho. After a small quake in early 1998, the average time between Old Faithful's eruptions increased a bit more, from about once every 76 minutes on average to about once every 80 minutes. Two geyser outbursts were 115 minutes apart. The 1998 quake temporarily reactivated some old geysers that had been quiet for years. Old Faithful expels water to an average height of 130 to 140 feet, but the range is between 98 and 183 feet.

Nearby and distant earthquakes change the timing and eruption height of geysers, probably by sealing or cracking open mineral deposits in a geyser basin's underground plumbing system, thus changing the flow of water into and out of the subterranean hot-water reservoirs that feed geysers. Seismic waves also can stretch and compress the ground, charging underground water pressures to change the geyser plumbing system. In geologic time, any particular geyser is a temporary feature. Geysers can become dormant when they are sealed by mineral deposits or by garbage discarded by vandals. Quakes can change their plumbing and divert water away from them or create openings from which new geysers erupt.

Cone-shaped mineral deposits are found at geysers and hot springs throughout Upper Geyser Basin. The cones are made of geyserite, also called siliceous sinter, which is a silica-rich mineral that precipitates out of the hot water.

As at West Thumb, the hot springs of Upper Geyser Basin are colored by various cyanobacteria. Biotechnology companies have been collecting these and other microbes from Yellowstone's springs for industrial and scientific uses and there has been debate over whether the government is receiving adequate compensation from such companies.

ON THE WAY TO STOP 3

After leaving Old Faithful parking lot, ramp merges onto main highway toward Madison: *1.2 miles from Stop 2; 6 miles from Stop 3.*

Black Sand Basin optional stop: *1.5 miles from Stop 2; 5.7 miles to Stop 3.* Colorful hot springs named Emerald Pool, Sunset Lake, and Rainbow Pool are located here, along with the unpredictable Cliff Geyser. The basin is named for black sand that formed from the erosion of obsidian, a black volcanic glass.

Biscuit Basin optional stop: *3.2 miles from Stop 2; 4 miles to Stop 3.* Biscuit Basin is a prime example of how earthquakes can change geysers and hot springs. Sapphire Pool became violently active, blew up, and destroyed its own plumbing after the 1959

Hebgen Lake earthquake. It no longer erupts. The basin has several other springs and geysers.

Turn left into Midway Geyser Basin parking lot: *7 miles from Stop 2; 0.2 miles to Stop 3.*

Stop 3: Midway Geyser Basin and Mallard Lake Dome

7.2 miles from Stop 2; 3.3 miles to Stop 4

This stop features Grand Prismatic Pool, the largest hot spring in North America and the deepest in Yellowstone at 111 feet, and Excelsior, a hot spring that once was a geyser but has been mostly inactive since the 1880s, when it erupted to heights of 50 to 300 feet. However, it erupted mildly for two days in September 1985 (Figure 9.4).

Like the Upper Geyser Basin, Midway Geyser Basin is located in a low, flat area left uncovered by surrounding lava flows. The lava-capped upward bulge named the Mallard Lake Dome is east of this stop. As it bulged upward due to molten rock under the Yellowstone caldera, two crack-like faults formed on top of the dome, similar to cracks forming on top of a loaf of baking bread. The faults run from southeast to northwest, and probably continue northwest beneath Midway Geyser Basin. The faults fracture rock to allow rain and snowmelt to percolate downward, become heated by hot rock, then rise upward along fault fractures that supply hot water to

9.4 ⁓ Midway Geyser Basin, with Grand Prismatic Spring (left) *and Excelsior Geyser* (right). *Grand Prismatic is a 200-foot-wide hot spring ringed by colorful mats of heat-resistant microbes. Two 400-foot-tall hills called Twin Buttes* (right background) *are made of glacial deposits and hot springs minerals. The ridge in the background is formed by lava flows along the western rim of the Yellowstone caldera.*

Midway's hydrothermal features. To the west is the Madison Plateau, capped by some of Yellowstone's more recent lava flows.

ON THE WAY TO STOP 4

Leave Midway Geyser Basin lot, turn left into highway: *0.2 miles from Stop 3; 3.1 miles to Stop 4.*

Leave main highway and turn east (right) onto Firehole Lake Drive: *1.1 miles from Stop 3; 2.2 miles to Stop 4.* After entering the Firehole Lake Drive—a one-way loop road—optional stops include Firehole Spring, Great Fountain Geyser, White Dome Geyser, and Pink Cone Geyser before reaching Steady Geyser and Firehole Lake. Great Fountain Geyser usually erupts twice daily and emits a massive fountain of water.

Stop 4: Steady Geyser and Firehole Lake

3.3 miles from Stop 3; 1.3 miles to Stop 5

Stop at the Steady Geyser-Firehole Lake parking lot, 2.2 miles after the entrance to Firehole Lake Drive.

Steady Geyser is a fountain geyser that erupts continuously through one of two vents, most often a 5-foot-high fountain from the lower vent. Firehole Lake is a large hot spring fed by several vents. In addition to the geyserite deposited by many of Yellowstone's geysers, the hydrothermal features in this area also deposit a whitish mineral named travertine.

ON THE WAY TO STOP 5

Continue on Firehole Lake Drive until it meets the main highway, then turn left. *1.1 miles from Stop 4; 0.2 miles from Stop 5.*

Turn right into Fountain Paint Pot parking lot: *1.2 miles from Stop 4; 0.1 miles from Stop 5.*

Stop 5: Fountain Paint Pot and Lower Geyser Basin

1.3 miles from Stop 4; 13.9 miles to Stop 6

Follow the trail past Silex Spring to Fountain Paint Pot and the geyser area.

Fountain Paint Pot provides visitors the best close-to-a-road view of mud pots in Yellowstone. Mud pots are found at the highest points in a hydrothermal basin. That puts them above the water table, so they contain only water from condensed steam or the ground surface. The water becomes acidic, breaking down rocks and minerals to form mud. Over many years, a muddy spring becomes a mud pot. Farther up the trail are Clepsydra, Fountain, Morning, and other geysers. Clepsydra erupts frequently.

ON THE WAY TO STOP 6

Leave Fountain Paint Pot parking lot, turn left onto highway: *0.1 miles from Stop 5; 13.8 miles to Stop 6.*

Caldera rim view: *1.2 miles from Stop 5; 12.7 miles to Stop 6.* The forested wall straight ahead in the distance is the northwest wall of the Yellowstone caldera. Unlike many other parts of the caldera wall and rim, this area has not been covered by lava flows since the caldera last exploded 630,000 years ago. What looks like the caldera wall is even more visible to the west (left). Unlike the north wall, however, what you are seeing is a lava flow that covered the west wall.

Turn left off main highway to enter Firehole Canyon bypass road: *7.6 miles from Stop 5; 6.3* miles *to Stop 6. Note:* You will pass the exit from this one-way road *before* you reach the entrance.

Lava flow: *8.1 miles from Stop 5; 5.8 miles to Stop 6.* You are driving through a lava flow. On your right, a cliff a few hundred feet high is visible across the Firehole River. It is a portion of a rhyolite lava flow. The river has cut down through the flow to reveal its internal structure. You can see contorted and jagged flow patterns and whorls from lava that flowed like toothpaste, twisting, turning, and even fracturing as it flowed like hot plastic. Named the West Yellowstone flow, this lava erupted about 111,000 years ago. More solidified lava is visible on your left (Figure 9.5) and right as you proceed up the road. In one spot you can see boulders within a flow, the result of hot lava encapsulating cooler chunks of lava that already had solidified.

Firehole Falls: *8.5 miles from Stop 5; 5.4 miles to Stop 6.* Falls formed here because the Firehole River reached the edge of a hard lava flow and eroded less resistant rock downstream.

Firehole River swimming holes, optional stop: *9.5 miles from Stop 5; 4.4 miles to Stop 6.* During hot weather, this is a great place to cool off in swimming holes carved out of solidified lava flows.

Exit Firehole Canyon bypass road, turn left on highway: *9.8 miles from Stop 5; 4.1 miles to Stop 6.* Note: You will re-enter the main highway above where

9.5 ⁓ *Lava flow along Firehole Canyon Drive includes angular rocks called breccias that formed as cooler, more brittle parts of the flowing lava broke into chunks. (Robert B. Smith)*

you entered the bypass road. Continue on the highway past the bypass road entrance.

Gibbon River crossing: *11.8 miles from Stop 5; 2.1 miles to Stop 6.* The idea of creating Yellowstone National Park was born at this point, near where the Gibbon and Firehole rivers meet to form the Madison River. Members of the Washburn–Langford–Doane expedition camped here in September 1870 and discussed preserving the area—an idea that gained momentum after Ferdinand V. Hayden's 1871 expedition. A bill designating Yellowstone as the first U.S. national park passed Congress and was signed by President Grant in 1872.

Turn left at Madison Junction: *12.2 miles from Stop 5; 1.7 miles to Stop 6.* Turn left at this intersection and proceed toward West Yellowstone, Montana. *Note an optional shortcut:* If you wish to omit Stops 6–12, including the tour of the Hebgen Lake earthquake area, *do not* turn left here. Instead, turn right and skip ahead to the Madison Junction section under "On the way to Stop 13."

Turn left into unmarked turnout across from Harlequin Lake sign on north (right): *13.8 miles from Stop 5; 0.1 miles from Stop 6.*

Stop 6: Madison Plateau View Turnout in Madison Canyon

13.9 miles from Stop 5; 12.3 miles to Stop 7

Just across from the Harlequin Lake trailhead sign on the north, this stop is one of several unmarked turnouts on the south side of the highway (which is left if you are heading toward West Yellowstone). Once you have pulled into the turnout, there is a sign reading Madison Plateau. However, other nearby turnouts are fine if you miss the correct one.

Deep in Madison Canyon, the northwest rim of the Yellowstone caldera is across the highway and above you to the north. This is one of the few places where it is possible to see part of the fault zone—the south-facing, mountainous wall north of the highway—created when the caldera floor dropped downward during its last gargantuan eruption about 630,000 years ago.

The Madison Plateau, above you to the south, is made of lava flows that filled in the caldera floor since the last caldera eruption. The Madison River eroded downward through old lava flows.

This area has experienced many earthquakes. During the magnitude-7.5 Hebgen Lake quake in 1959, large rocks tumbled off Mount Jackson and blocked the highway.

The area is also notable for earthquake swarms, which are sequences of hundreds or thousands of small quakes without a much stronger foreshock or mainshock.

During late 1985, a nearby seismic swarm produced hundreds of small quakes daily. The strongest was about magnitude 4.5. The quakes were centered on faults that begin inside the caldera southeast of here, then run northwest through this area and toward the fault zone that produced the magnitude-7.5 Hebgen Lake quake in 1959. At about the same time, the floor of the Yellowstone caldera started sinking after having bulged upward since measurements began in 1923. It is possible the quakes were caused by hot water or molten rock flowing outward from beneath the caldera along the faults.

ON THE WAY TO STOP 7

Exit turnout, turn left (west) on highway: *0.1 miles from Stop 6; 12.2 miles to Stop 7.*

Leaving the Yellowstone Plateau: *4.5 miles from Stop 6; 7.8 miles to Stop 7.* As a highway bridge crosses the Madison River, you are emerging from the lava flows of the Yellowstone Plateau and entering the broad, flat West Yellowstone Basin, which measures roughly 10 by 15 miles. This basin was smoothed and flattened by a giant glacier flowing from the icecap that repeatedly covered the Yellowstone Plateau. The glacier scoured the basin and filled it with sand and gravel eroded from lava flows in Yellowstone. As you enter West Yellowstone Basin, note the elongated, hummocky hills on both sides of the road. They were formed as flowing glacial ice gouged the landscape.

Leave Yellowstone National Park at west entrance: *11.9 miles from Stop 6; 0.4 miles to Stop 7.* Proceed west to West Yellowstone, Montana. Stop 7, the information center, is on the left.

Stop 7: West Yellowstone Chamber of Commerce Information Center; Start Tour to Hebgen Lake Quake Area

12.3 miles from Stop 6; 11.6 miles to Stop 8

Rest rooms, information, and phones are available at the information center. Restaurants, motels, and tourist attractions are nearby. You have a few options here:

1. End the first day of your Yellowstone tour here and stay overnight. Start the next morning with the tour of the Hebgen Lake earthquake area (Stops 8–11),

then stay here again tomorrow night. Then proceed on the rest of the tour the next day. That breaks this chapter into a three-day tour.

2. Stay here overnight, do the Hebgen Lake tour early tomorrow morning, then proceed back into Yellowstone for the rest of the tour. That makes this chapter a two-day tour, with a long second day.

3. If you started early this morning and arrive early afternoon, you still may have time to proceed with Stops 8 through 11, the tour of the Hebgen Lake earthquake area, then return to West Yellowstone (Stop 12) for the night before proceeding on a second and final day of the tour. However, if you leave too late, the Hebgen Lake Earthquake Visitor Center may be closed.

ON THE WAY TO STOP 8

If you wish to skip the tour of the Hebgen Lake earthquake area, jump from Stop 7, West Yellowstone, to Stop 12, which also is West Yellowstone.

If you wish to take the optional tour of the Hebgen Lake earthquake area, drive north from West Yellowstone on U.S. 191–287 toward Bozeman, Montana.

Duck Creek Y, turn left on U.S. 287 toward Ennis, Montana: *8.6 miles from Stop 7; 3 miles to Stop 8.* This is the gateway to Hebgen Lake and Earthquake Lake. Some claim that the quake's epicenter is near here, but other studies suggest it was located about 2 miles to the northwest.

Stop 8: Hebgen Lake Earthquake Area Sign Turnout

11.6 miles from Stop 7; 9.4 miles to Stop 9 (14.8 miles if you take optional side trip to Red Canyon)

This stop is near the east end of two fault segments that broke the night of August 17, 1959. The magnitude-7.5 Hebgen Lake earthquake killed twenty-eight people, most in a gigantic landslide 17 miles down the Madison River from where you stand. (See Chapter 1 and Figures 1.1 and 1.2 for more about the disaster.)

During the quake, the land under Hebgen Lake dropped downward and tilted northward, permanently inundating docks and some lakeshore homes on this side of the lake. Land north of the highway rose upward.

Those opposing movements broke the ground to create two major "scarps," which are short, steep banks where the fault breaks the ground surface. The two scarps are cracks in the ground many miles long and roughly parallel. Both are segments or branches of the Hebgen Lake fault. One is the 8-mile-long Hebgen segment, which begins less than 2 miles west of here, then pretty much parallels the lake's north shore. It will be seen at Stop 10. The other is the 14-mile-long Red Canyon fault segment, which runs through the hills a couple hundred yards north of where you stand. It is a 10-foot bank, but is covered with vegetation and difficult for an untrained eye to detect. However, a bit west of here, the fault turns up Red Canyon and runs into the mountains before turning west again. This part of the Red Canyon fault segment is visible from an interpretive sign just down the highway.

A "normal" fault is the kind of fault in which land on one side of the fault line rises and ground on the other side drops down and away, like a block sliding down a ramp. The Hebgen Lake earthquake was the most powerful normal-fault earthquake in historic time in the United States. There have been bigger quakes, but those were on other kinds of faults.

ON THE WAY TO STOP 9

Red Canyon scarp interpretive sign: *1.1 miles from Stop 8; 8.3 miles to Stop 9 (13.7 miles if you take optional trip to Red Canyon).* If the weather is clear, the Red Canyon fault scarp is visible as a gash across the base of the mountain in the distance up the canyon north (right) of the road. As you drive onward from this point, look north to see more of the scarp, still evident decades after the quake.

Optional side trip up Red Canyon Road and mile round-trip hike to fault scarp: *Optional right turn into Red Canyon Road is 1.7 miles from Stop 8 and 7.7 miles to Stop 9. If you take this 5.4-mile side trip, the distance to Stop 9 is 13.1 miles.* To reach the trailhead, turn right onto Red Canyon Road at the sign marked Red Canyon, which is just past milepost 18 on U.S. 287. Bear right again at another sign marked Red Canyon. The trailhead is 2.7 miles from Highway 287, at the end of unpaved Red Canyon Road near a large boulder, an outhouse, and a sign for the Cabin Creek–Red Canyon Trail No. 205.

Hike about one-half mile up the dirt road, which dwindles to a trail as it goes up canyon. You will switchback once down canyon, then turn up canyon on a second switchback. Walk about 100 yards past the second switchback to the scarp. It is the dirt cliff or bank in front of you. You should be standing in a side ravine to the main canyon. The ravine crosses the trail. Look up the ravine to see a more impressive face of the scarp.

This is the largest scarp created by the 1959 earthquake. Here the ground on one side of the fault rose while the other side dropped, for net movement of 22 feet.

Imagine if the ground broke apart like this in a city. The only reason the Hebgen Lake quake did not claim many more lives was that it struck a sparsely populated area.

Turn around, return to the highway, and turn west (right).

From U.S. 287, turn left into Building Destruction area: *9.2 miles from Stop 8 (14.6 miles if you took side trip to Red Canyon); 0.2 miles to Stop 9.*

Stop 9: Building Destruction from Hebgen Lake Quake

9.4 miles from Stop 8 (14.8 miles if you took optional side trip to Red Canyon); 1.7 miles to Stop 10

Walk down a short trail to see what is left of old cabins ravaged by the 1959 quake. Large waves sloshed back and forth across the lake for more than 11 hours after the quake. Two buildings were wrecked here and a third was dropped into the lake during the quake. A woman and dog who occupied it escaped the water and survived. In addition to the waves, land beneath the lake dropped 19 feet downward and tilted northward due to fault movements during the quake. This permanently shifted the lake northward, inundating the north shore of the lake and some buildings on it. Meanwhile, the land rose and the waterline permanently dropped on the south shore, leaving docks and boats high and dry.

ON THE WAY TO STOP 10

Exit Building Destruction area, turn left on highway: *0.2 miles from Stop 9; 1.5 miles to Stop 10.*

Hebgen Lake Dam: *1.1 miles from Stop 9; 0.6 miles to Stop 10.* Hebgen Lake Dam is 87 feet tall. The dam and its spillway were damaged, but didn't fail during the 1959 quake despite extremely strong shaking that sent large waves sloshing over the dam. It took almost 12 hours for the sloshing to die down.

Stop 10: Fault Scarp at Cabin Creek

1.7 miles from Stop 9; 5.8 miles to Stop 11

Turn right into Cabin Creek fault scarp area. (It is the turn just before you cross Cabin Creek and reach the Cabin Creek Campground.) Park on the left. Walk up the short trail past the picnic table. The 20-foot-high dirt bank in front of you is the Hebgen Lake fault scarp created during the 1959 quake. The ground on which you stand dropped downward and the ground atop the bank rose upward during the quake, literally splitting the old Cabin Creek campground in half and temporarily trapping some campers.

ON THE WAY TO STOP 11

Continue on loop road at Cabin Creek scarp area until it reaches highway, then turn right: *0.2 miles from Stop 10; 5.6 miles to Stop 11.*

Refuge Point: *0.9 miles from Stop 10; 4.9 miles to Stop 11.* The 1959 quake unleashed the gigantic landslide almost 5 miles down canyon from here and wrecked the highway near Hebgen Lake about 5 miles upstream from this point. That trapped about 250 campers in the Madison River Canyon. Many of them, including some who were injured, made their way to higher ground near Refuge Point. Forest Service smokejumpers parachuted to this point the morning after the quake, providing first aid. By afternoon, helicopters landed to evacuate the injured and stranded.

Earthquake Lake sign turnout: *3.1 miles from Stop 10; 2.7 miles to Stop 11.* The massive landslide, still almost 3 miles downstream from here, dammed the Madison River, creating Earthquake Lake. Decades after the quake, the haunting snags of old, inundated trees still protrude from the lake.

Turn right onto driveway for Hebgen Lake Earthquake Visitors Center: *5.5 miles from Stop 10; 0.3 miles to Stop 11.*

Stop 11: Hebgen Lake Earthquake Visitor Center

5.8 miles from Stop 10; 28 miles to Stop 12

The visitor center's displays tell the story of the 1959 earthquake and the disastrous landslide in the Madison River Canyon. The center is well worth a visit. The real display, however, is the landslide-denuded mountainside outside the visitor center. (See Figure 1.1.) As you stand in this quiet spot, it is difficult to envision the violence that once occurred.

The Hebgen Lake earthquake triggered the collapse of the entire mountainside across the Madison River from the visitor center, leaving the slope that remains mostly barren decades later. The slide crossed the riverbed and rose 450 feet up this side of the canyon. The visitor center was built on slide debris. The force of the slide created a large wave of water, as well as winds so strong that they ripped the clothes off victims, including some survivors.

Nineteen of the quake's twenty-eight known victims died here, and remain entombed beneath 80 million tons of debris. Seven others suffered fatal injuries or died when hit by rocks, washed away by a 30-foot-tall, slide-induced wave or drowned as river water rapidly backed up behind the slide. The other two were crushed in their tent by a falling boulder at Cliff Lake Campground, 7 miles southwest. Walk or drive uphill to Memorial Boulder, a 3,000-ton rock carried across the canyon by the slide. The boulder bears a plaque with the names of the dead.

ON THE WAY TO STOP 12

Turn around: From the Hebgen Lake Earthquake Visitor Center return to West Yellowstone, Montana.

Stop 12: West Yellowstone Chamber of Commerce Information Center

28 miles from Stop 11; 27.8 miles to Stop 13

Here is a place to stay overnight or take a break before proceeding on the final leg of the Yellowstone tour.

ON THE WAY TO STOP 13

West entrance Yellowstone National Park: *0.4 miles from Stop 12; 27.4 miles to Stop 13.*

Turn north (left) at Madison Junction: *14 miles from Stop 12; 13.8 miles to Stop 13.* As you drive up Madison Canyon from Madison Junction to Gibbon Falls, you are paralleling the northwest wall of the caldera, visible to the north (left). Lava flows that erupted after the last major caldera explosion are to the south (right). Older flows are to the north (left).

Gibbon River Canyon and Gibbon Falls: *18.8 miles from Stop 12; 9 miles to Stop 13.* This is an unmarked turnout on the east (right) less than a half mile past the Gibbon River picnic area. A lava flow is exposed in the rock wall across the river. The falls, visible below you, formed where rock eroded more slowly than less resistant rock downstream.

Gibbon Meadow: *23.3 miles from Stop 12; 4.5 miles to Stop 13.* As the road rises out of Gibbon Canyon, you enter large, open meadows located between volcanic flows. To the northwest (left) are high peaks of the Gallatin Range. They are ancient mountains that once extended farther south until that part of the range was destroyed by Yellowstone's first caldera explosion 2 million years ago.

At Norris Junction, turn west (left) into Norris Geyser Basin parking lot: *27.3 miles from Stop 12; 0.5 miles to Stop 13.*

Stop 13: Norris Geyser Basin

27.8 miles from Stop 12; 21 miles to Stop 14

Norris Geyser Basin has Yellowstone's hottest hydrothermal reservoir, with drill hole instruments measuring a record 459 degrees Fahrenheit about 1,000 feet beneath the basin. Many springs and geysers emit boiling water. Norris is also the park's oldest and most unstable geyser basin. The basin has two main loop trails:

1. The Back Basin Trail passes Echinus Geyser, which erupts 40 to 60 feet high every 40 to 80 minutes and is the world's most acidic geyser—comparable to

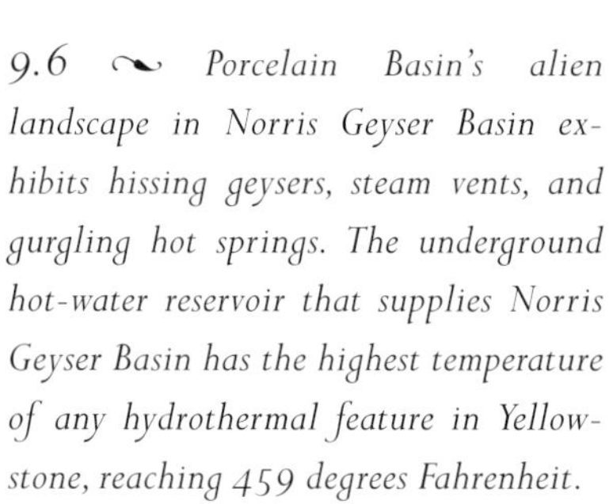

9.6 ~ Porcelain Basin's alien landscape in Norris Geyser Basin exhibits hissing geysers, steam vents, and gurgling hot springs. The underground hot-water reservoir that supplies Norris Geyser Basin has the highest temperature of any hydrothermal feature in Yellowstone, reaching 459 degrees Fahrenheit.

vinegar. The world's tallest active geyser, Steamboat Geyser, is also on this trail (see Figure 4.4). It can erupt to heights of almost 400 feet, but goes years without doing so. More often, it hurls water 40 feet or less.

2. The Porcelain Basin Trail takes you to the most intensely active part of Norris, a large area of hot springs and geysers that also can be observed from the overlook at the Norris Museum (Figure 9.6).

Minerals deposited from the hot water clog Norris's underground plumbing system, beginning an annual cycle that ends in a dramatic disturbance of the geysers and hot springs. Pressure builds as the hot-water system becomes sealed. Winter snowfall and the water table act as lids to help contain the pressure. In summer, the snow is gone and the water table drops, reducing containment pressure. Critical pressure is reached during summer in drier years and during fall in wetter years. An underground eruption or explosion occurs as the mineral seals clogging the pipes are broken suddenly. In a matter of hours, all geysers and springs in the basin become muddy and turbid. There is also a big surge in hot-water and steam discharges. Some geysers

erupt suddenly after months of dormancy. Eruption intervals change for other geysers. Steamboat dries up and just emits steam. Echinus Geyser becomes erratic.

Sometimes Norris's instability can be quite violent, producing above-ground hydrothermal explosions. One night in 1971, such a blast blew out a 40-foot chunk of Porcelain Basin, scattering rocks and debris. Porkchop Geyser was once a quiet hot spring and occasional geyser, then started spewing water continuously in 1985, then tripled in height, and finally exploded in 1989, hurling rocks and chunks of geyserite nearly 220 feet. Some rocks landed at the feet of startled onlookers.

Norris also is noted for intense swarms of small earthquakes—and some larger ones. A magnitude-6.1 quake was centered near Norris Junction in March 1975, causing extensive damage to park roads and facilities. It was the second-largest quake in Yellowstone's recorded history, after the Hebgen Lake quake and its largest aftershocks.

Seismic and geothermal activity at Norris is related to the intersection of various faults and volcanic vents. Several faults—including the East Gallatin fault and a possible extension of the Hebgen Lake fault—appear to extend to Norris. The basin also sits near the Yellowstone caldera's northern rim, which is mostly buried by lava flows. When the caldera last blew up, its floor dropped downward along a fault that forms a huge ring around the inside of the caldera rim. Norris sits just north of that caldera ring fault. Volcanic eruption vents and lines of quake epicenters stretch from southeast to northwest within the caldera and extend to Norris. The intersection of faults, volcanic vents, and the caldera boundary create zones of weakness within Earth's crust, making it easier for heat to rise from molten and partly molten rock beneath the caldera. That heats water in the ground, and the hot water moves through faults—both upward and northward—to create the geysers and hot springs at Norris and Mammoth Hot Springs (Stop 14). Earthquakes help preserve the geysers and hot springs by shaking loose the mineral deposits that regularly seal hot-water conduits.

ON THE WAY TO STOP 14

Leave Norris Geyser Basin parking lot and turn north (left): *0.4 miles from Stop 13; 20.6 miles to Stop 14.* Proceed toward Mammoth Hot Springs. (*Note:* If you are short of time and wish to skip Mammoth and other sites in Stops 14–16, go straight instead of left and proceed to Canyon Junction, located in the section On the way to Stop 17. Turn right when you reach Canyon Junction.)

Roaring Mountain turnout: *5.3 miles from Stop 13; 15.7 miles to Stop 14.* An unusual hydrothermal system climbs Roaring Mountain on the east (right) side of the road. Geysers no longer are active here, perhaps due to inadequate heat in the area or clogging of hot-water conduits by mineral deposits. Fumaroles or steam vents remain active. Acid in the steam bleached rocks near the vents, creating the area's gray, ashy appearance. Roaring Mountain gets its name from steam emissions that can range from inaudible to a hiss to a roar loud enough to be heard a few miles away.

Obsidian Cliff driveby: *8.7 miles from Stop 13; 12.3 miles to Stop 14.* The cliff is on your right, but you cannot stop because old turnouts have been barricaded by the Park Service due to repeated theft of glass-like, black obsidian rock from the cliff. The cliff is part of a 180,000-year-old lava flow that cooled in such a way to form the large columns at the base of the cliff. Obsidian is a lava that solidified into glass because it cooled too quickly for crystals to form. An interpretive sign turnout is on the left about 0.1 miles past the cliff.

Swan Lake and view of Gallatin Range: *15.2 miles from Stop 13; 5.8 miles to Stop 14.* As you enter the flat area around Swan Lake, the Gallatin Range is visible farther to the west (left). A north–south fault like the Teton fault—but not as active—runs along the east base of the Gallatin Range. The mountains rose along the fault. To the east (right) is Bunsen Peak. It is the neck or plug of solidified magma that rose within a 50-million-year-old volcano. Most of the volcano eroded away, leaving Bunsen Peak.

Golden Gate Cliffs turnout on right: *16.9 miles from Stop 13; 4.1 miles to Stop 14.* The cliffs come into view as you begin to drop off the Yellowstone Plateau. The cliffs are welded ash-flow tuffs, which are rocks that formed from dense flows of hot ash blown out of the Yellowstone caldera by the big eruption 2 million years ago.

View of the Hoodoos, leave the Yellowstone Plateau: *17.5 miles from Stop 13; 3.5 miles to Stop 14.* You are now dropping off the Yellowstone Plateau. The Hoodoos, which are unusual rock formations, represent a large landslide of travertine mineral that originated in a large hydrothermal system like the one downslope at Mammoth Hot Springs. The old hot springs deposited the travertine and later died. Then rocks and mineral deposits slid from the mountainside above on the west (left) and continued down toward the valley to the east (right).

Mammoth Terrace Drive: *19.3 miles from Stop 13; 1.7 miles to Stop 14.* (An optional detour onto this one-way loop will add about 1.5 miles to the distance to Stop 14.) The Upper Terrace of Mammoth Hot Springs is a geothermal system with extensive travertine deposits that form large white terraces.

Entering Mammoth Hot Springs sign: *20.7 miles from Stop 13; 0.3 miles to Stop 14.*

Stop 14: Mammoth Hot Springs, Lower Terraces

21 miles from 13; 20.1 miles to Stop 15

This stop is the parking lot on the west (left) 0.3 miles past the Entering Mammoth Hot Springs sign. You also can park just past it in the lot for Liberty Cap.

Mammoth's springs owe their existence to presence of a zone of faults extending from here south to Norris Geyser Basin. The faulting allows hot fluids below the caldera to move north and feed Mammoth Hot Springs. The terraces exist because acidic hot water percolates upward through limestone bedrock, which was deposited when ancient seas covered the area. When hot springs water erupts to the surface and cools, limestone precipitates in the form of another mineral, travertine, which forms the terraces.

Highlights of the Lower Terraces include the multicolored microbial growth in Palette Spring and travertine deposits in Minerva Terrace and Opal Terrace (Figure 9.7), which has gobbled up a tennis court and threatened a historic house. Liberty Cap, more than 36 feet tall, was formed by minerals deposited by a hot spring that is now inactive. Liberty Cap was named in 1871 by Ferdinand V. Hayden's expedition because it looked like hats worn during the French Revolution.

ON THE WAY TO STOP 15

Mammoth Hot Springs Visitor Center: *0.4 miles past Stop 14; 19.7 miles to Stop 15.* The visitor center includes exhibits and rest rooms. You also may wish to see the Mammoth Hotel nearby.

Take road toward Tower Junction: If you have not stopped at the visitor center, it is the right turn just before the visitor center lot. If you did stop, turn left out of the visitor center lot, then immediately left again.

Gardiner River glaciation: *1.3 miles from Stop 14; 18.8 miles to Stop 15.* A glacier once flowed off the Yellowstone Plateau and down the Gardiner River Valley, depositing debris to form the hummocky topography around you. As you drive the next several miles, edging around the north base of the Yellowstone Plateau, look at the sloping hills and ice-carved valleys and imagine how glaciers flowed north–northwest off the plateau and into this area as recently as 14,000 years ago.

9.7 ～ *Opal Terrace at Mammoth Hot Springs is composed of calcium-rich travertine. The terrace was created by mineral-laden hot water spreading out across a steep landscape, depositing minerals as step-like platforms. (Rick Hutchinson.)*

Fossil tree optional side trip: *Turn right 17.1 miles from Stop 14; rejoin main road 3 miles to Stop 15.* (This optional side trip will add 1.1 miles to the distances in this section.) Turn right down this half-mile road to see a petrified redwood tree buried and preserved by mud flows from pre-Yellowstone volcanic eruptions that built the Absaroka Range about 50 million years ago. There is a forest of such fossil trees on Specimen Ridge east of Tower Junction.

Proceed straight through Tower Junction (toward Canyon Junction): *18.6 miles from Stop 14 (excluding side trip); 1.5 miles to Stop 15.*

Stop 15: Calcite Springs Overlook

20.1 miles from Stop 14; 13.4 miles to Stop 16

From the turnout on the east (left) side of the road, walk the left branch of the short loop trail up to the overlook of the river and springs, then down the stairs to the viewpoint looking into the Yellowstone River canyon.

Calcite Springs are located on the pale slope that is near river level downstream. The rock has been whitened by chemicals from the springs, which are warmed by heat rising from an underground volcanic fracture zone. There are deposits of oil and other hydrocarbons in the rocks beneath the springs. The rising heat acts like a natural oil refinery, driving oils out of deeper sedimentary rocks and making them ooze to the surface at Calcite Springs. Southeast of here, Rainbow Hot Springs also has naturally occurring oil and hydrocarbons.

From the lower part of the overlook, you can see how the Yellowstone River has cut deeply through rock layers to create The Narrows (Figure 9.8), which provide a look back through geologic time:

9.8 ~ *The Narrows of the lower Grand Canyon of the Yellowstone exemplify how canyons reveal geologic history. At the bottom of the gorge are 50-million-year-old volcanic rocks from eruptions that formed the Absaroka Range. Columnar basalts, stream deposits, and volcanic ash form layers higher in the canyon.*

- The rocks making up most of the vertical height of the canyon, including the oddly eroded rocks right across from you, are coarse sands, gravels, and cobbles deposited by ancient glaciers and the Yellowstone River.
- Above those rocks, and closer to the top of the canyon, is rock from a basalt lava flow that cooled to create vertical columns within the flow. The lava flowed in a 25-foot-deep flood that covered this part of Yellowstone during eruptions about 1.3 million years ago.
- Sitting atop the basalt (and covered by grass and trees), are volcanic ash deposits from the Yellowstone caldera explosion 630,000 years ago. Glacial sediments sit above the ash on the surface.

ON THE WAY TO STOP 16

Optional walk to Tower Fall: *0.8 miles from Stop 15; 12.6 miles to Stop 16.* Restrooms and food are available here. Walk a short paved path to the waterfall view. Tower Fall flows over a cliff of volcanic rock from eruptions of Absaroka volcanoes 50 million years ago. Glacial rock deposits, which are less resistant, eroded to form the waterfall.

The Presence of Grizzlies turnout: *5.5 miles from Stop 15; 7.9 miles to Stop 16.* This open, grassy terrain is prime habitat for grizzly bears, but once was covered by glacial ice. To the south, topped by a fire lookout, is 10,243-foot Mount Washburn, the remnant of an old volcano that erupted as part of the activity that built that Absaroka Range 50 million years ago. Mount Washburn is just outside the north edge of the Yellowstone caldera.

Dunraven Pass and optional hike to the summit of Mount Washburn: *11.9 miles from Stop 15; 1.5 miles to Stop 16.* Dunraven Pass is near the northern rim of the Yellowstone caldera. Shortly after you cross the pass, you will be looking into the caldera created by the huge eruption 630,000 years ago. You may take an optional 6-mile-round-trip hike (with a 1,300-foot vertical gain) to the 10,243-foot summit of Mount Washburn, the highest peak in north–central Yellowstone. The summit overlooks the Yellowstone Plateau and caldera to the south.

Stop 16: Heart of the Caldera Sign Turnout

13.4 miles from Stop 15; 7.2 miles to Stop 17

Stop at the turnout on the east (left) with the sign reading Heart of the Caldera. Be careful to watch for oncoming traffic.

You now are standing on the northern edge of the Yellowstone caldera. This stop provides spectacular views of several features:

- To the south, forests cover the mostly flat terrain of the center of the Yellowstone Plateau between Old Faithful and the meadows of Hayden Valley. This area is the heart of the giant crater known as the Yellowstone caldera. It is now relatively flat because it was covered by about thirty large lava flows that erupted since the caldera's last catastrophic eruption 630,000 years ago. Standing here, you get a sense of the immense size of the caldera, extending roughly 45 miles from northeast to southwest and about 30 miles from the northwest to the southeast. Imagine the entire area in front of you suddenly dropping a few hundred yards downward during a massive caldera eruption as hot magma explodes past you at supersonic speeds and ash billows miles skyward. If you were standing here during such an eruption, you would be vaporized. You actually are looking at parts of two overlapping calderas. The central portion of Yellowstone is the caldera that blew up most recently, some 630,000 years ago. It overlaps with the now-obscured caldera, located more to the southwest, that exploded 2 million years ago in Yellowstone's first huge caldera eruption. (The caldera that blew up 1.3 million years ago is located farther west at Island Park, Idaho.)

- To the south–southeast (left), past the trees, are the open meadows of Hayden Valley, which once was the bottom of ancestral Yellowstone Lake. You also can see part of the modern lake. The mountain behind Hayden Valley in the distance is Mount Sheridan, located near the south edge of the caldera. To the west (right) of Mount Sheridan, the Teton Range is visible on clear days.

- Behind Hayden Valley but in front of Mount Sheridan is a forested ridge named Elephant Back. Beneath this ridge is a fault zone that runs between the Sour Creek and Mallard Lake resurgent domes, allowing underground hot water and magma to flow between them. The two domes have been bulging up and down since the last big caldera eruption. Molten and partly molten rock that power Yellowstone's geothermal features is underground beneath the domes. The Sour Creek dome is visible here as a forested hill southeast (left) of Elephant Back.

- Below you to the east, Washburn Hot Springs are located in the bare, gray-colored areas surrounded by forests. The springs exist because hot water rises along fault lines that encircle the ancient caldera floor.

- To the east and southeast, the distant mountains are the peaks of the Absaroka Range. To the east, but closer, are reddish rocks of the west-facing wall of the Grand Canyon of the Yellowstone River.

ON THE WAY TO STOP 17

Proceed straight through Canyon Junction: *3.3 miles from Stop 16; 3.9 miles to Stop 17.* (*Note:* If you want a rest stop, do not proceed straight. Instead, turn east [left] and drive 0.2 miles before turning right into to the Canyon Visitor Center parking lot. After your rest stop, turn left out of the lot, return 0.2 miles to Canyon Junction, turn south [left] and proceed toward Stop 17.)

Turn east (left) on South Rim Drive at sign for Grand Canyon of the Yellowstone—Artist Point: *5.6 miles from Stop 16 (or 2.3 miles from Canyon Junction); 1.6 miles to Stop 17.* Proceed 1.6 miles to the parking lot for Stop 17, Artist Point.

Stop 17: Artist Point, Grand Canyon of the Yellowstone

7.2 miles from Stop 16; 8.9 miles to Stop 18

Of several spectacular overlooks on the north and south rims of the Grand Canyon of the Yellowstone River, Artist Point provides resplendent views of the canyon and 308-foot-tall Lower Falls (see Figure 6.2), which formed because resistant rhyolite lava rock did not erode as much as lava that had been weakened by heat and chemicals in an ancient hot springs basin. A similar process created the 109-foot-tall Upper Falls, out of sight upstream.

The rocks in the canyon walls are rhyolites from a lava eruption 275,000 years ago. The Yellowstone River cut downward through the solidified lava flows to form this canyon. As downcutting occurs, the falls move backward to the south, increasing the length of the canyon. The Grand Canyon of the Yellowstone extends 20 miles from the canyon area north toward Tower Junction.

During the magnitude-6.1 Norris earthquake of 1975, centered 15 miles west of here, some of the sidewalks at this viewpoint were destroyed. With sharp drop-offs into the canyon, some people feared for their lives and fled the area.

ON THE WAY TO THE STOP 18

Exit South Rim Drive and turn south (left) onto highway toward Yellowstone Lake: *1.6 miles from Stop 17; 7.3 miles to Stop 18.*

9.9 ⁓ *Rich grasslands in Hayden Valley provide feed for bison grazing near Trout Creek. The grass grows on sediments deposited when ancient Yellowstone Lake covered the area. (Rick Hutchinson)*

Enter Hayden Valley, wildlife viewing: *3.5 miles from Stop 17; 5.4 miles to Stop 18.* For about 6 miles, you will drive through the open, grassy meadows of Hayden Valley, where you might encounter bison, wolves, and bears—and traffic jams of tourists watching them (Figure 9.9). Drive slowly. The lava flows that filled the Yellowstone caldera didn't quite reach Hayden Valley, creating a natural basin. The Yellowstone icecap—a 3,500-foot-thick layer of ice—sat atop this area during several glacial periods over the past 2 million years. The icecap both carved out the landscape and deposited piles of rocky debris, making the topography hummocky. Later, this area was covered by a large ancestor to Yellowstone Lake. It was almost 300 feet deeper than the modern lake, thanks to melting glaciers and a dam of glacial ice that blocked the Yellowstone River near the Grand Canyon of the Yellowstone. Thick sediments deposited at the lake bottom smoothed the hummocky landscape, creating relatively flat terrain in Hayden Valley. If the ancient lake was still here—instead of dammed several miles upstream—Hayden Valley would be under a few hundred feet of water. The ice dam eventually melted and broke, and the lake level dropped significantly.

Hayden Valley Former Lakebed interpretive sign: *8.4 miles from Stop 17; 0.5 miles to Stop 18.* Old lake sediments fostered grasses and shrubs that feed bison and other animals.

Sulfur Caldron: *8.8 miles from Stop 17; 0.1 miles to Stop 18.* Parking on east (left). Yellowish because of its high concentration of sulfur, this spring has an acidity as strong as battery acid.

Stop 18: Mud Volcano

8.9 miles from Stop 17; 3 miles to Stop 19

Parking area is on the right. Trails and boardwalks take visitors past several mud pots and steam vents (Figure 9.10).

Only steam and other gases, not hot water, reach Mud Volcano from underground hydrothermal reservoirs. The only water in Mud Volcano and nearby mud pots comes from the condensation of steam and from shallow groundwater from rain and snowmelt.

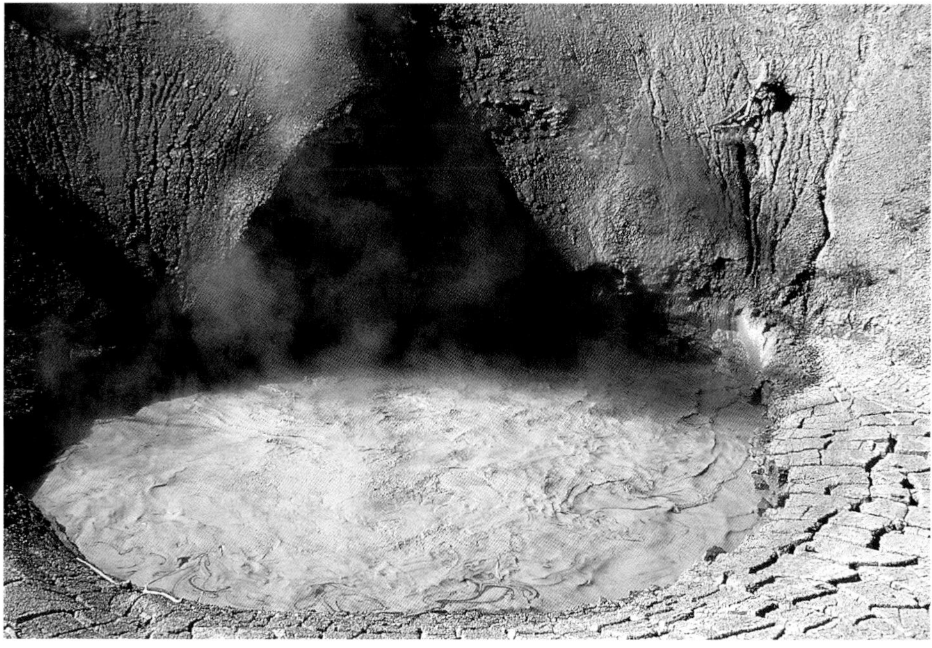

9.10 ~ Mud Volcano in Hayden Valley is famous for thundering noise produced by steam issuing from the throat of the spring. Dark mud, formed by the decomposition of surrounding rock and decaying bacteria, spatters onto the walls, blackening them. The pungent smell of hydrogen sulfide gas is apparent at Mud Volcano.

The area is similar to the Geysers geothermal field in California, where you can drill several hundred feet underground and still hit only steam, not hot water. Mud Caldron, Mud Volcano, and other mud pots are muddy because hydrogen sulfide gas (the rotten-egg smell), water, and bacterial growth help create sulfuric acid that breaks down rocks. Without much flowing water, it is impossible to flush out the mud. Steam bubbles up through the mud so it looks like it is boiling, even though it is not.

Hayden Valley and the Mud Volcano area are close to the point of maximum historic uplift and sinking of the caldera floor, which is heavily faulted here. Many small swarms of quakes occurred here, including some that triggered geysers across the highway. A 1978–1979 swarm allowed more heat to flow upward into the area named Cooking Hillside. The heat killed trees

Mud Volcano is located here partly because various faults intersect in this area, creating zones of weakness within the ground. This permits water to percolate deep underground where it is heated to form steam, which then rises to the surface. During the 1978 swarm, small geysers erupted for a few months through the Yellowstone River across the highway.

Stop 19: Le Hardy Rapids Turnout (Unmarked)

3 miles from Stop 18; 13.2 miles to Stop 20

This unmarked turnout is on the northeast (left) side of the highway. Walk the boardwalk to the Yellowstone River to see the rapids.

Le Hardy Rapids are not particularly spectacular rapids, but they are significant for what they tell us about Yellowstone's volcanism and earthquakes and the water level in modern Yellowstone Lake.

At these rapids, the Yellowstone River runs east to west and crosses a series of faults that run southwest to northeast. Land downstream from the faults dropped downward relative to land on the upstream side, which has been lifted upward, creating the stairstep-like series of rapids in volcanic rocks that are highly resistant to erosion. This uplift has helped dam the Yellowstone River, which begins to pond upstream from the rapids, allowing water to back up and impound Yellowstone Lake at Fishing Bridge, a few miles south. Much of Yellowstone Lake sits within a basin created by the collapse of the caldera floor during the last huge eruption 630,000 years ago. That basin was scoured out more recently by glaciers. But the presence of the faults and the rapids has raised the lake level higher than it would be otherwise.

The faults that create Le Hardy Rapids run southwest to northeast between the two resurgent domes within the Yellowstone caldera. The faults start at the Mallard Lake Dome near Old Faithful, run northeast under the ridge named Elephant Back Mountain, then cross under the Yellowstone River at Le Hardy Rapids and continue to Sour Creek Dome, the timbered bluff located north across the river. The faults formed as hot water and molten rock beneath the caldera made the entire caldera—and the two domes—rise upward and drop downward over time.

Only a half-mile from this stop is a surveyor's benchmark where the greatest upward movement of the caldera was measured between 1923 and 1984. During that time, the caldera floor was lifted more than 3 feet vertically. From 1985 to the mid-1990s, it dropped downward again by 6 to 8 inches, then began rising again.

The faults created by the up-and-down "breathing" of the volcanic caldera not only formed Le Hardy Rapids and helped impound modern Yellowstone Lake, but also generated numerous earthquakes in this area.

ON THE WAY TO STOP 20

Turn east (left) to Fishing Bridge: *2.9 miles from Stop 19; 10.3 miles to Stop 20.*

Fishing Bridge: *3.2 miles from Stop 19; 10 miles to Stop 20.* Fishing Bridge, located at the north end of Yellowstone Lake, is the lake's outlet. After crossing the bridge, the highway traverses an old beach that was lifted above the present lake level as molten rock under the caldera made the caldera floor bulge upward. Some researchers have estimated the ground beneath the north end of the lake has risen roughly 20 or 30 yards in the last 14,000 years. While the lake's north end has bulged upward, ground beneath the lake's southern end tilted downward, submerging trees and other vegetation along the south shore.

Indian Pond: *6.2 miles from Stop 19; 7 miles to Stop 20.* Indian Pond, on the south (right), formed in the crater blown out by a shallow steam and hot-water eruption several thousand years ago.

Mary Bay: *7.1 miles from Stop 19; 6.1 miles to Stop 20.* Mary Bay is an elongated bay on the northeast end of Yellowstone Lake. It is another explosion crater that formed when a steam eruption blew a hole in the ground. The explosion blasted out sand, gravel, and rocks. The debris fell all along the lake's north shore. Measurements of the bottom of Mary Bay show a large hole and water temperatures near boiling.

Steamboat Point: *9.4 miles from Stop 19; 3.8 miles to Stop 20.* Steamboat Point is the large rocky peninsula sticking westward into Yellowstone Lake. As you drive around the point, the rock in the road cut and cliff east (left) of the highway is breccia, a rock containing other rocks, which have sharp rather than smooth edges. The breccia was deposited when Mary Bay exploded, scattering rock fragments that later were cemented into new bedrock. Once you have crossed Steamboat Point, Sedge Bay is to the west (right). Like Mary Bay, the lake bed beneath Sedge Bay has hot areas and may have been formed by a steam explosion.

Turn left onto road to Lake Butte Overlook: *12.4 miles from Stop 19; 0.8 miles to Stop 20.* Proceed 0.8 miles to the overlook.

Stop 20: Lake Butte Overlook and End of Tour

13.2 miles from Stop 19; end of tour

If the weather is clear, your tour ends with a glorious view of Yellowstone Lake from this overlook, located on the eastern edge of the Yellowstone caldera.

Lake Butte is volcanic rock created by the eruptions that formed the Absaroka Range about 50 million years ago. It might have been a hill cut in two as the caldera

floor dropped downward during the last major eruption 630,000 years ago. If such an eruption occurred now, anyone standing on Lake Butte would be incinerated instantly.

Also imagine what it would be like here during the Ice Age. The point where you are standing would be buried beneath an icecap some 3,500 feet thick. The ice helped carve out the basin for Yellowstone Lake—a job started by the last caldera explosion.

The view to the south encompasses the Southeast and South arms of the lake. The caldera boundary runs to Lake Butte from the northeast, then cuts roughly southwest across the lake. So most of the southeast portion of the lake is outside the giant crater. The basin occupied by that part of the lake was carved out by glaciers and also formed by faulting along the Southeast and South arms. The rest of the lake occupies only the eastern edge of the 45-by-30-mile caldera.

The deepest portion of Yellowstone Lake—about 300 feet deep—is located to the southwest roughly between where you are standing and Frank Island. Another underwater hot-water system is on the lake bottom just east of Stevenson Island, the island west of here.

To the southwest, the high mountain is 10,308-foot Mount Sheridan, located near the southern edge of the Yellowstone caldera. On a clear day, you may see the Teton

9.11 ~ *View of Yellowstone Lake from Lake Butte, with Sedge Bay* (right foreground) *separated from Mary Bay* (right background) *by Steamboat Point. The high ridge in the background is part of Elephant Back Mountain.*

Range farther southwest on the skyline and Yellowstone Lake's West Thumb to the west–southwest. As you take in the view from the southwest to the west, you can get a sense of the caldera's topography: A crater that was filled in with lava flows to form today's flattened Yellowstone Plateau surrounded by higher mountains.

To the northwest are Sedge and Mary bays (Figure 9.11), formed by steam eruption holes in the ground a mile or two wide.

Internet Sites

AUTHOR ROBERT B. SMITH'S WEB SITE FOR YELLOWSTONE AND GRAND TETON RESEARCH

http://www.mines.utah.edu/~rbsmith/rbs-home.index.html

EARTHQUAKES

Global earthquake monitor:
http://www.iris.washington.edu/seismic/60_2040_1_8.html

Global and U.S. earthquake information:
http://wwwneic.cr.usgs.gov/

Surfing the web for earthquake information:
http://www.geophys.washington.edu/seismobig.html

Yellowstone earthquakes:
http://www.seis.utah.edu/HTML/YPSeismicityMaps.html

Grand Teton National Park earthquakes:
http://www.seismo.usbr.gov/seismo/eqdata.html#worldwide

U.S. seismic hazard maps:
http://geohazards.cr.usgs.gov/eq/

Global list of earthquake and volcano web sites:
http://www-socal.wr.usgs.gov/seismolinks.html

VOLCANOES AND GEYSERS

Global volcano monitor:
http://volcano.und.nodak.edu/vwdocs/current_volcs/current.html

Global volcano information:
http://www.nmnh.si.edu/gvp/

Surfing the web for volcano information:
http://eost.u-strasbg.fr/~hugues/subvolcano.html

U.S. volcano information:
http://volcanoes.usgs.gov/

Yellowstone volcano information:
http://vulcan.wr.usgs.gov/Volcanoes/Yellowstone/framework.html

Yellowstone Geysers
http://www.geocities.com/Yosemite/1407/geyser_main.html

General volcano hazards:
http://volcanoes.usgs.gov/Hazards/What/hazards.html

NATIONAL PARK SERVICE HOMEPAGES

Grand Teton National Park:
http://www.nps.gov/grte/

Yellowstone National Park:
http://www.nps.gov/yell/

GENERAL PARK INFORMATION

The Great Outdoor Recreation Page for Grand Teton National Park:
http://www.gorp.com/gorp/resource/US_National_Park/wy_grand.html

Grand Teton National Park page:
http://www.grand.teton.national-park.com/

The Great Outdoor Recreation Page for Yellowstone National Park:
http://www.gorp.com/gorp/resource/US_National_Park/wy_yello.html

The Total Yellowstone Page:
http://www.yellowstone-natl-park.com/

Natural Resources Bibliography for Yellowstone and Grand Teton National Parks
http://www.nature.nps.gov/nrbib/

EDUCATIONAL WEB SITES ON PLATE TECTONICS, EARTHQUAKES, AND VOLCANOES

Plate tectonics:
http://pubs.usgs.gov/publications/text/dynamic.html

The Savage Earth:
http://www.pbs.org/wnet/savageearth/

National Geographic Fantastic Journeys:
http://www.nationalgeographic.com/yellowstone/

The Jason Project at Yellowstone:
http://www.jasonproject.org/expeditions/jason8/yellowstone/

Earthquakes, plate tectonics and space methods in Earth sciences:
http://scign.jpl.nasa.gov/learn/

Investigating earthquakes:
http://www.scecdc.scec.org/Module/module.html

Earth Alert, daily updates on the state of the planet:
http://www.discovery.com/news/earthalert/earthalert.html

References

Chapter 1—A Land of Scenery and Violence

Christopherson, E. *The Night the Mountain Fell: The Story of the Montana-Yellowstone Earthquake.* Missoula, Montana: Earthquake Press, 1960.

Clary, David A. *The Place Where Hell Bubbled Up: A History of the First National Park.* Moose, Wyoming: Homestead Publishing, 1993.

Dunn, I. B. *Out of the Night.* Sandpoint, Idaho: Plaudit Press, Keokee Co. Publishing Inc., 1998.

Jaggar, T. A. "A plea for geophysical and geochemical observatories." *J. Washington Aca. Sciences.* 12: 343–53, 1922.

Nauta, D., and J. C. Janetski. *Yellowstone Earthquake! The Story of the 1959 Hebgen Lake Disaster* (videotape). West Yellowstone, Montana: Yellowstone Publications, 1997.

O'Brien, E., and S. J. Nava. *1959 Hebgen Lake, Montana Earthquake—Newspaper Articles.* (News stories from August 1959 from the *Billings Gazette, Bozeman Daily Chronicle, Montana Standard, The Salt Lake Tribune, The Associated Press, United Press International,* and *The Deseret News.*) Salt Lake City: Earthquake Education Services, University of Utah Seismograph Stations, 1997.

Smith, R. B., and R. L. Christiansen. "Yellowstone Park as a window on the Earth's interior." *Scientific American.* 242: 104–17, 1980.

U.S. Forest Service, Gallatin National Forest. Displays, video, and interpretive signs at Hebgen Lake Earthquake Visitor Center, Montana.

Wicks Jr., C., W. Thatcher, and D. Dzurisin. "Migration of fluids beneath Yellowstone Caldera inferred from satellite radar interferometry." *Science.* 282: 458–64, 1998.

Witkind, I. J. *The Night the Earth Shook: A Guide to the Madison River Canyon Earthquake Area.* U.S. Department of Agriculture, U.S. Forest Service Miscellaneous Publication No. 907. Washington, D.C.: U.S. Government Printing Office, 1962: O-643789.

Chapter 2—In the Wake of the Yellowstone Hotspot

Alvarado, G. E., P. Denyer, and C. W. Sinton. "The 89 Ma Tortugal komatiitic suite, Costa Rica: Implications for a common geological origin of the Caribbean and Eastern Pacific region from a mantle plume." *Geology.* 25: 439–42, 1997.

Anders, M. H., J. W. Geissman, L. A. Piety, and J. T. Sullivan. "Parabolic distribution of circum-eastern Snake River Plain seismicity and latest Quaternary faulting—migratory pattern and association with the Yellowstone hotspot." *J. Geophys. Res.* 94: 1589–1621, 1989.

Armstrong, R. L., W. P. Leeman, and H. E. Malde. "K-Ar dating, Quaternary and Neogene volcanic rocks of the Snake River Plain Idaho." *Amer. J. Sci.* 275: 225–51, 1975.

Braile, L. W., R. B. Smith, J. Ansorge, M. R. Baker, M. A. Sparlin, C. Prodehl, M. M. Schilly, J. H. Healy, S. T. Mueller, and K. H. Olsen. "The Yellowstone–Snake River Plain seismic profiling experiment: Crustal structure of the eastern Snake River Plain." *J. Geophys. Res.* 87: 2597–2609, 1982.

Camp, V. E. "Mid-Miocene propagation of the Yellowstone mantle plume head beneath the Columbia River basalt source region." *Geology*. 23: 435–38, 1995.

Chapman, M., and R.L. Kirk. "A migratory mantle plume on Venus: Implications for Earth?" *J. Geophys. Res.* 101: 15953–67, 1996.

Duncan, R. A., and M. A. Richards. "Hotspots, mantle plumes, flood basalts and true polar wander." *Reviews of Geophysics.* 29: 31–51, 1991.

George, R., N. Rogers, and S. Kelley. "Earliest magmatism in Ethiopia: Evidence for two mantle plumes in one flood basalt province." *Geology*. 26: 923–26, 1998.

Good, J. M., and K. L. Pierce. *Interpreting the Landscapes of Grand Teton and Yellowstone National Parks: Recent and Ongoing Geology*. Moose, Wyoming: Grand Teton Natural History Association, 1996.

Harder, H., and U. E. Christensen. "A one-plume model of martian mantle convection." *Nature*. 380: 507–09, 1996.

Humphreys, E. D., and K. G. Dueker. "Physical state of the western U. S. upper mantle." *J. Geophys. Res.* 99: 9625–50, 1994.

Johnston, S. T., P. J. Wynne, D. Francis, C.J. R. Hart, R.J. Enkin, and D. C. Engebretson. "Yellowstone in the Yukon: The Late Cretaceous Carmacks Group." *Geology*. 24: 997–1000, 1996.

Kerr, R. A. "A deep root for Iceland?" *Science.* 279: 806, 1998.

Kerr, R. A. "Fiery Io models Earth's first days." *Science*. 280: 381–82, 1998.

Knight-Ridder News Service. "Jovian satellite appears a likely habitat for E.T." *The Salt Lake Tribune,* October 19, 1998.

Leeman, W. P. "Development of the Snake River Plain–Yellowstone Plateau Province, Idaho and Wyoming: An overview and petrologic model." *Bull. Idaho Bur. Mines and Geol.* 26: 155–78, 1982.

Lithgow-Bertelloni, C., and P. G. Silver. "Dynamic topography, plate driving forces and the African superswell." *Nature*. 395: 269–72, 1998.

Marty, B., B. G. J. Upton, and R. M. Ellam. "Helium isotopes in early Tertiary basalts, northeast Greenland: Evidence for 58 Ma plume activity in the Northern Atlantic–Iceland volcanic province." *Geology*. 26: 407–10, 1998.

McEwen, A. S., L. Keszthelyi, J. R. Spencer, G. Schubert, D. L. Matson, R. Lopes-Gautier, K. P. Klaasen, T. V. Johnson, J. W. Head, P. Geissler, S. Fagents, A. G. Davies, M. H. Carr, H. H. Breneman, and M. J. S. Belton. "High-temperature silicate volcanism on Jupiter's moon Io." *Science*. 281: 87–90, 1998.

Murphy, J. B., G. L. Oppliger, G. H. Brimhall Jr., and A. Hynes. "Plume-modified orogeny: An example from the western United States." *Geology.* 26: 731–34, 1998.

National Aeronautics and Space Administration. "Searching for Yellowstone on Mars." NASA Ames Research Center news release 96-58, October 28, 1996.

Perkins, M. E., W. P. Nash, F. H. Brown, and R. J. Fleck. "Fallout tuffs of Trapper Creek, Idaho–A record of Miocene explosive volcanism in the Snake River Plain." *Bull. Geol. Soc. Amer.* 107: 1484–1506, 1995.

Pierce, K. L., and L.A. Morgan. "The track of the Yellowstone hot spot: Volcanism, faulting, and uplift." *Geol. Soc. Amer. Memoir.* 179: 1–53, 1992.

Saltzer, R., and E. Humphreys. "Upper mantle P wave velocity structure of the eastern Snake River plain and its relationship to geodynamic models of the region." *J. Geophys. Res.* 102: 11829–41, 1997.

Schilly, M. M., R. B. Smith, J. Ansorge, J. A. Lehman, and L. W. Braile. "The Yellowstone–eastern Snake River Plain seismic profiling experiment: Upper-crustal structure." *J. Geophys. Res.* 87: 2692–2704, 1982.

Self, S., Th. Thordarson, L. Keszthelyi, G. P. L. Walker, K. Hon, M. Y. Murphy, P. Long, and S. Finnemore. "A new model for the emplacement of Columbia River basalts as large, inflated pahoehoe lava flow fields." *Geophys. Res. Letters.* 23: 2689–92, 1996.

Smith, R. B. "Intraplate tectonics of the Western North American Plate." *Tectonophysics.* 37: 323–36, 1977.

Smith, R. B., and L. W. Braile. "Topographic signature, space-time evolution, and physical properties of the Yellowstone-Snake River Plain volcanic system: The Yellowstone hotspot," in Snoke, A.W., J. R. Steidtmann, and S. M. Roberts. *Geology of Wyoming: Geological Survey of Wyoming Memoir No. 5*, 1993, pp. 694–754.

Smith, R. B., and L. W. Braile. "The Yellowstone hotspot." *J. Volcan. and Geothermal Res.* 61: 121–88, 1994

Smith, R. B., L. W. Braile, M. M. Schilly, J. Ansorge, C. Prodehl, M. Baker, J. H. Healey, S. Mueller, and R. Greensfelder. "The Yellowstone–eastern Snake River Plain seismic profiling experiment: Crustal structure of Yellowstone." *J. Geophys. Res.* 87: 2597–2609, 1982.

Smith, R. B., and M. L. Sbar. "Contemporary tectonics and seismicity of the western United States, with emphasis on the Intermountain Seismic Belt." *Bull. Geol. Soc. Amer.* 85: 1205–18, 1974.

Smith, R. B., R. T. Shuey, R. Freidline, R. Otis, and L. Alley "Yellowstone hot spot: New magmatic and seismic evidence." *Geology.* 2: 451–55, 1974.

University of California, Santa Cruz. "Roots of 'hot spots' may extend to Earth's core-mantle boundary." *UC Santa Cruz Tip Sheet*, February 1997.

Wolfe, C. J., I.Th. Bjarnason, J. C. VanDecar, and S. C. Solomon. "Seismic structure of the Iceland mantle plume." *Nature.* 385: 245–47, 1997.

Chapter 3—Cataclysm! The Hotspot Reaches Yellowstone

Christiansen, R. L. "Yellowstone magmatic evolution: Its bearing on understanding large-volume explosive volcanism." *Explosive Volcanism, Its Inception, Evolution and Hazards.* Washington, D.C.: National Academy of Sciences, 1984, pp. 84–95.

Christiansen, R. L. "The Quaternary and Pliocene Yellowstone volcanic field of Wyoming, Idaho and Montana." U.S. Geological Survey Professional Paper 729-G, in press.

Christiansen, R. L., and H. R. Blank Jr. "Volcanic stratigraphy of the Quaternary rhyolite plateau in Yellowstone National Park." U.S. Geological Survey Professional Paper 729-B, 1972.

Clary, David A. *The Place Where Hell Bubbled Up: A History of the First National Park.* Moose, Wyoming: Homestead Publishing, 1993.

DeLucas, Kathy. "First detailed images of New Mexico volcano show that Valles Caldera still underlain by magmatic melt" (news release). Los Alamos, N.M.: Los Alamos National Laboratory, December 16, 1996.

Dzurisin, D., J. C. Savage, and R. O. Fournier. "Recent crustal subsidence at Yellowstone caldera." *Bull. Volcan.* 52: 247–70, 1990.

Hayden, F. V. "Report." Preliminary report of the United States Geological Survey of Montana and Portions of Adjacent Territories, U.S. Geological and Geographical Survey of the Territories Fifth Annual Report [for 1871], 1872, pp. 13–204.

Hayden, F. V. "Report." Sixth Annual Report of the United States Geological Survey in the Territories, U.S. Geological and Geographical Survey of the Territories Sixth Annual Report [for 1872], 1873, pp. 13–85.

Heiken, Grant. Los Alamos National Laboratory, personal communication, Dec. 2, 1998.

Hill, David. U.S. Geological Survey, personal communication, June 1998.

Hutchinson, R. A. "Geologic publications and articles related to Yellowstone National Park." Yellowstone National Park, Wyoming: National Park Service, Wyoming Yellowstone Center for Resources, YCR-NR-97-3, 1997.

Keefer, W. R. "The geologic story of Yellowstone National Park." *U. S. Geol. Survey Bull.* 1347, 1975.

Muffler, L. J. P., D. E. White, and A. H. Truesdale. "Hydrothermal explosion craters in Yellowstone National Park." *Bull. Geol. Soc. Amer.* 82: 723–40, 1971.

Sarna-Wojcicki, A.M., and J. O. Davis "Quaternary tephrochronology." *Quaternary Nonglacial Geology: Conterminous U.S., The Geology of North Amer.* K-2: 93–116, 1991.

Smith, R. B., J. O. D. Byrd, and D. D. Susong. "The Teton fault: Seismotectonics, Quaternary history and earthquake hazards." Geological Survey of Wyoming Memoir No. 5, 628–67, 1993.

Chapter 4—How Yellowstone Works

The Associated Press (Idaho–Wyoming state wire). "Changes in geyser eruptions show up after January earthquake," June 22, 1998.

The Associated Press. "LSD? You bet; Nobel winner promises open mind as witness," April 5, 1995.

The Associated Press. "U.S. scientists win physics; Chemistry for Canadian, American" and "Excerpts from Nobel chemistry prize announcement," October 13, 1993.

Christiansen, R. L. "The Quaternary and Pliocene Yellowstone volcanic field of Wyoming, Idaho, and Montana." U.S. Geological Survey Professional Paper 729-G, in press.

Clary, David A. *The Place Where Hell Bubbled Up: A History of the First National Park.* Moose, Wyoming: Homestead Publishing, 1993.

Dzurisin, D., J. C. Savage, and R. O. Fournier. "Recent crustal subsidence at Yellowstone caldera." *Bull. Volcan.* 52: 247–70, 1990.

Dzurisin, D., K. Yamashita, and J. W Kleinman. "Mechanisms of crustal uplift and subsidence at the Yellowstone caldera."*Bull. Volcan.* 56: 261–70, 1994.

Fournier, R. O." Geochemistry and dynamics of the Yellowstone National Park hydrothermal system." *Ann. Rev. Earth Planet. Sci.* 17: 13–53, 1989.

Gowans, F. R. *A Fur Trade History of Yellowstone Park: Notes, Documents, Maps.* Orem, Utah: Mountain Grizzly Publications, 1989.

Holdahl, S. R., and D. Dzurisin. "Time dependent models of vertical deformation for the Yellowstone–Hebgen Lake region, 1923–1987." *J. Geophys. Res.* 96: 2465–83, 1991.

Hutchinson, R. A., J. A. Westphal, and S. W. Kieffer. "In situ observations of Old Faithful Geyser." *Geology.* 25: 875–78, 1997.

Janetsky, J. C. *Indians of Yellowstone Park.* Salt Lake City: Bonneville Books, University of Utah Press, 1987.

Locke, W. M., and G. A. Meyer. "A 12,000-year record of vertical deformation across the Yellowstone caldera margin: The shorelines of Yellowstone Lake." *J. Geophys. Res.* 99: 20079–94, 1994.

Meertens, C. M., R. B. Smith, and D. W. Vasco. "Kinematics of crustal deformation of the Yellowstone hotspot using GPS." *Eos, Trans. Amer. Geophys. Un.* 74, Supplement 63, 1993.

Miller, D. S., and R. B. Smith. "P and S velocity structure of the Yellowstone volcanic field from local earthquake and controlled source tomography." *J. Geophys. Res.* In press.

Morgan, P., D. D. Blackwell, R. E. Spafford, and R. B. Smith. "Heat flow measurements in Yellowstone Lake and the thermal structure of the Yellowstone caldera." *J. Geophys. Res.* 82: 3719–32, 1977.

New York Times. "Yellowstone's Microbial Riches Lure Eager Bioprospectors," Section F, page 3, column 1, October 14, 1997.

Pelton, J. R., and R. B. Smith. "Contemporary vertical surface displacements in Yellowstone National Park." *J. Geophys. Res.* 87: 2745–61, 1982.

Pitt, A. M. "Map showing earthquake epicenters (1964–81) in Yellowstone National Park and vicinity, Wyoming, Idaho, and Montana." U.S. Geological Survey Miscellaneous Field Studies Map MF-2022, 1989.

Savage, J. C., M. Lisowski, W. H. Prescott, and A. M., Pitt. "Deformation from 1973 to 1987 in the epicentral area of the 1959 Hebgen Lake, Montana, earthquake ($M_S = 7.5$)" *J. Geophys. Res.* 898: 2145–54, 1992.

Smith, R. B., and W. J. Arabasz. "Seismicity of the Intermountain Seismic Belt." *Geol. Soc. Amer.* SMV V-1, Decade Map Volume 1, 185–228, 1991.

Smith, R. B., and L. W. Braile. "Crustal structure and evolution of an explosive silicic volcanic system at Yellowstone National Park." *Explosive Volcanism, Its Inception, Evolution and Hazards.* Washington, D. C.: National Academy of Sciences, 1984, pp. 96–109.

Smith, R. B., and M. L. Sbar. "Contemporary tectonics and seismicity of the western United States, with emphasis on the Intermountain Seismic Belt." *Bull. Geol. Soc. Amer.* 85: 1205–18, 1974.

White, D. E., R. A Hutchinson, and T. E. C. Keith. "The geology and remarkable thermal activity of Norris Geyser Basin, Yellowstone National Park, Wyoming." U.S. Geological Survey Professional Paper 1456, 1988.

Wicks Jr., C., W. Thatcher, and D. Dzurisin. "Migration of fluids beneath Yellowstone Caldera inferred from satellite radar interferometry." *Science*. 282: 458–64, 1998.

Chapter 5—The Broken Earth: Why the Tetons Are Grand

Byrd, J. O. D., R. B. Smith, and J. W. Geissman. "The Teton fault, Wyoming: Topographic signature, neotectonics and mechanism of deformation." *J. Geophys. Res.* 99: 20095–122, 1994.

Case, J. C. *Earthquakes and Active Faults in Wyoming.* Laramie: Geological Survey of Wyoming, 1991.

Case, J. C. *Earthquakes and Related Geologic Hazards in Wyoming.* Geological Survey of Wyoming, Public Information Circular 26, 1986.

Fryxell, F. M. *The Tetons: Interpretations of a Mountain Landscape*. Berkeley: University of California Press, 1938.

Hodgson, Bryan. "Grand Teton." *National Geographic*. 187: 119–40, 1995.

Love, J. D., and J. C. Reed Jr. *Creation of the Teton Landscape, Teton National Park*. Moose, Wyoming: Grand Teton Natural History Association, 1971.

Mattes, M. J. *Jackson Hole, Crossroads of the Western Fur Trade,* Jackson, Wyoming: Jackson Hole Museum and Teton County Historical Society, 1994.

Morris, G. A. and P. R. Hooper. "Petrogenesis of the Colville Igneous Complex, northeast Washington: Implications for Eocene tectonics in the northern U.S. Cordillera." *Geology*. 25(9): 831–34, 1997.

Righter, R. W. *Crucible for Conservation: A Struggle for Grand Teton National Park*. Boulder: Colorado Associated University Press, 1982.

Smith, R. B., J. O. D. Byrd, and D. D. Susong. "The Teton fault: Seismotectonics, Quaternary history and earthquake hazards." *Geological Survey of Wyoming Memoir.* 5: 628–67, 1993.

Smith, R. B., J. R. Pelton, and J. D. Love, "Seismicity and the possibility of earthquake-related landslides in the Teton–Gros Ventre–Jackson Hole area, Wyoming." *Contrib. to Wyoming Geol.* 14: 57–64, 1976.

U. S. Geological Survey, *Grand Teton National Park, Wyoming* (topographic map), Washington, D.C. 1968.

Wood, C. "Seismicity of the Teton region, Wyoming and Idaho; Perspective from a newly installed telemetered network." *1988 Abstracts with Program, Rocky Mountain Section, Geol. Soc. Am.* 20: A14, 1988.

Chapter 6—Ice over Fire: Glaciers Carve the Landscape

The Associated Press. News stories on Loki volcano, October 2, October 5, October 9, and November 5, 1996.

The Associated Press. "Scientists speculate on Mount Rainier eruption," December 12, 1980.

Bargar, K. E., and R. O. Fournier. "Effects of glacial ice on subsurface temperatures of hydrothermal systems in Yellowstone National Park, Wyoming: Fluid-inclusion evidence." *Geology*. 16: 1077–80, 1988.

Einarsson, P., B. Brandsdottir, M.T. Gudmundsson, H. Bjornsson, and K. Grinvold. "Center of the Iceland Hotspot Experiences Volcanic Unrest." *Eos, Trans. Amer. Geophys. Un.* 78(35): 369 and 374–75, 1997.

Gosse, J. C., J. Klein, E. B. Evenson, B. Lawn, and R. Middleton. "Beryllium-10 dating of the duration and retreat of the Last Pinedale glacial sequence." *Science*. 268: 1329–33, 1995.

Locke, W. M. "An equilibrium Yellowstone icecap, in Late Pleistocene–Holocene Evolution of the Northeastern Yellowstone Landscape." *Late Pleistocene-Holocene Evolution of the Northeastern Yellowstone Landscape, 1995 Field Conference Guidebook.* Friends of the Pleistocene, Rocky Mountain Cell, 1, 1995.

Muffler, L. J. P., D. E. White, and A. H. Truesdale. "Hydrothermal explosion craters in Yellowstone National Park." *Bull. Geol. Soc. Amer.* 82: 723–40, 1971.

Otis, R. M., R. B. Smith, and R. J. Wold. "Geophysical surveys of Yellowstone Lake, Wyoming." *J. Geophys. Res.* 82: 3705–18, 1977.

Pierce, K. L. "History and dynamics of glaciation in the northern Yellowstone National Park area." U.S. Geological Survey Professional Paper 729F, 1979.

Pierce, K. L., and J. Good "Quaternary geology of Jackson Hole, Wyoming."Geological Survey of Wyoming, Public Information Circular 29: 79-88, 1990.

Richmond, G. M. "Stratigraphy and chronology of glaciations in Yellowstone National Park." *Report of the International Geological Correlation Programme Project 24.* Oxford, England: Pergamon Press, 1984, p. 83.

Smith, R. B., K. L. Pierce, and R. J. Wold. "Quaternary history and structure of Jackson Lake, Wyoming, from seismic reflection surveys." *Geological Survey of Wyoming Memoir.* 5: 668–93, 1993.

Chapter 7—Future Disasters

The Associated Press. "Dream on," an AP graphic published in *The Salt Lake Tribune,* May 20, 1998.

The Associated Press. News stories about Pinatubo during 1991 and 1992.

The Associated Press. Volcano forecast, October 24, 1991.

Bonney, O., and L. Bonney. *Battle Drums and Geysers.* Chicago: Sage Books, Swallow Press, 1970.

Briffa, K. R., P. D. Jones, F. H. Schweingruber, and T. J. Osborn. "Influence of volcanic eruptions on Northern Hemisphere summer temperature over the past 600 years." *Nature.* 393: 450–56, 1998.

Case, J. C. *Earthquakes and Active Faults in Wyoming.* Laramie: Geological Survey of Wyoming, 1991.

Case, J.C. "Heavy precipitation could increase landslide hazards in Wyoming" (news release). Laramie: Geological Survey of Wyoming, February 4, 1997.

Eddy, J. A., "Changes in time of the temperature of the Earth." *Earthquest*. 5: insert, 1997.

Fisher, R. V., G. Heiken, and J. B. Hulen. *Volcanoes: Crucibles of Change*. Princeton, New Jersey: Princeton University Press, 1997.

Francis, P. *Volcanoes, a Planetary Perspective*. Oxford, England: Oxford University Press, 1996.

Gilbert, J. D., D. Ostenna, and C. Wood. "Seismotectonic study, Jackson Lake Dam and Reservoir." Minidoka Project, Idaho-Wyoming, U.S. Bureau of Reclamation Seismotectonic Report 83-8, 1983.

Hill, David P. Personal communication, June 1998.

Nebraska Game and Parks Commission. *Ashfall Fossil Beds, A State Historical Park* (brochure). 1998.

Newhall, C. G., and R. S. Punongbayan, editors. *Fire and Mud: Eruptions and Lahars of Mount Pinatubo, Philippines*. Quezon City: The Philippine Institute of Volcanology and Seismology; Seattle and London: University of Washington Press, 1996.

Perkins, M. E., and W. P. Nash. "Temporal variation in explosive silicic volcanism along the Yellowstone hotspot track: The fallout tuff record." *Bull. Geol. Soc. Amer.* In press.

Reese, R. *Greater Yellowstone, The National Park and Adjacent Wildlands*. Montana Geographic Series, No. 6, Second Edition, Helena, Montana: Montana Magazine, American and World Geographic Publishing, 1991.

Sarna-Wojcicki, A.M. Personal communication, 1995.

Shipley, S., and A. M. Sarna-Wojcicki. "Distribution, thickness, and mass of Late Pleistocene and Holocene tephra from major volcanoes in the northwestern United States: A preliminary assessment of the hazards from volcanic ejecta to nuclear reactors in the Pacific northwest." U. S. Geological Survey Miscellaneous Field Studies Map, MF-1435, 1983.

Siegel, Lee. Personal notes of U.S. Geological Survey seismic hazards mapping workshop, Salt Lake City, Utah, February 16, 1995.

Stommel, H., and E. Stommel. "The year without a summer." *Scientific American*. 240: 176–83, 1979.

Teton County Municipal Emergency Operations Plan, 1985.

U.S. Bureau of Reclamation. Personal communications with Diana Cross, public affairs officer, Boise, Idaho, December 1997.

U.S. Bureau of Reclamation. *Standard Operating Procedures for Jackson Lake Dam*. March 1984.

Varley, J. D. "Rock to fins: The unlikely connections between geology and aquatic life in Yellowstone Park." Unpublished report for the May 1998 Long Term Ecological Research planning meeting, Bozeman, Montana, 1998.

Zreda, M., and J. S. Noller. "Ages of prehistoric earthquakes revealed by cosmogenic chlorine-36 in a bedrock fault scarp at Hebgen Lake." *Science*. 282: 1097–99, 1998.

Chapter 8—Grand Teton Tour

National Park Service, U.S. Department of the Interior. *Grand Teton Official Map and Guide*. Washington, D.C.: U.S. Government Printing Office, 1995—387-038/00286 Reprint 1995.

Smith, R. B., J. O. D. Byrd, and D. D. Susong. "Neotectonics and structural evolution of the Teton fault." *Geologic field tours of western Wyoming and parts of adjacent Idaho, Montana, and Utah.* Geological Survey of Wyoming, Public Information Circular 29: 126–38, 1990.

Trails Illustrated Topo Maps, map 202, *Grand Teton National Park, Wyoming.* Evergreen, Colorado: Ponderosa Publishing Co., 1994.

Chapter 9—Yellowstone Tour

The Associated Press (Idaho–Wyoming state wire). "Changes in geyser eruptions show up after January earthquake," June 22, 1998.

Fournier, R. O., R. L. Christiansen, R. A. Hutchinson, and K. L. Pierce. "Yellowstone National Park field trip: Volcanic, hydrothermal, and glacial activity in the Yellowstone region." *U. S. Geol. Surv. Bull.* 2099, 1994.

Fritz, William J. *Roadside Geology of the Yellowstone Country.* Missoula, Montana: Mountain Press Publishing Co., March 1996.

Gallatin National Forest. *Madison River Canyon Earthquake Area Tour Guide.* West Yellowstone, Montana: U.S. Forest Service, 1997.

Hutchinson, R.A. Personal communication, 1996.

National Park Service, U.S. Department of the Interior, *Yellowstone Official Map and Guide.* Washington, D.C.: U.S. Government Printing Office, 1996—404-952/40209 Reprint 1996.

Trails Illustrated Topo Maps, map 201, *Yellowstone National Park, Wyoming.* Evergreen, Colorado: Ponderosa Publishing Co., 1994.

U.S. Geological Survey, Department of the Interior, *Ferdinand Vandiveer Hayden and the Founding of Yellowstone National Park.* Washington, D.C.: U.S. Government Printing Office, 1980—311-348-1.

The Yellowstone Association. *West Thumb Geyser Basin; Old Faithful, Upper Geyser, Black Sand and Biscuit Basins; Fountain Paint Pot and Firehole Lake Drive; Norris Geyser Basin; Mammoth Hot Springs; Canyon; and Mud Volcano* (brochures). Yellowstone National Park, Wyoming: The Yellowstone Association, 1996.

Index

Note: Page numbers in bold indicate illustrations.